3rd International Workshop on Metamaterial-By-Design 2017

AA006315

Journal of Physics: Conference Series Volume 963

Madrid, Spain
14 - 15 December 2017

Editors:

Giacomo Oliveri
Eva Rajo-Iglesias

ISBN: 978-1-5108-5895-4
ISSN: 1742-6588

Printed from e-media with permission by:

Curran Associates, Inc.
57 Morehouse Lane
Red Hook, NY 12571

CURRAN ASSOCIATES INC. proceedings.com

Some format issues inherent in the e-media version may also appear in this print version.

Copyright© (2017) by the Institute of Physics
All rights reserved. The material featured in this book is subject to
IOP copyright protection, unless otherwise indicated.

Printed by Curran Associates, Inc. (2018)

For permission requests, please contact the Institute of Physics
at the address below.

Institute of Physics
Dirac House, Temple Back
Bristol BS1 6BE UK

Phone: 44 1 17 929 7481
Fax: 44 1 17 920 0979

techtracking@iop.org

Additional copies of this publication are available from:

Curran Associates, Inc.
57 Morehouse Lane
Red Hook, NY 12571 USA
Phone: 845-758-0400
Fax: 845-758-2633
Email: curran@proceedings.com
Web: www.proceedings.com

Table of contents

Volume 963

International Workshop on Metamaterial-By-Design
14–15 December 2017, Madrid, Spain

Accepted papers received: 1 February 2018
Published online: 21 February 2018

Preface

Papers

IOP Proceedings on
2017 International Workshop on Metamaterial-by-Design
Preface

The International Workshop on Metamaterials-by-Design (http://www.2017-metamaterials-workshop.eu/) has been established since 2015 within a very active research area in ICT, material science, engineering, and physics. In fact, the possibility to engineer materials able to exhibit the desired effective properties is an emerging research topic in several areas, because of the envisaged possibility to reduce the cost, enhance the features and flexibility, and obtain behaviours not achievable with standard media simply by combining (through suitably mixtures/ordered arrangements) common materials, which has gathered the interest of a vast community of engineers, scientists, chemists, and physics with strong interdisciplinary skills.

Despite these promising features, the research in new generation artificial materials (usually collectively indicated as Metamaterials) is still in its infancy from the scientific viewpoint, and several important challenges need to be addressed to bridge the gap between academic prototypes and consumer products, specifically in the field of Communications and Sensing.

In this framework, "Metamaterials-by-Design" (MbD) has emerged as a powerful paradigm to address the above challenges. *MbD* is an instance of the System-by-Design paradigm defined in short as "How to deal with complexity". More specifically, *MbD* considers the problem of designing artificial-material enhanced-devices from a completely new perspective, that is

The application-oriented synthesis of advanced systems comprising artificial materials whose constituent properties are driven by the device functional requirements

This paradigm is based on the concept that any device comprising artificial materials must be designed (*i*) considering its functionality as the fundamental objective for the synthesis, and that (*ii*) the properties of the media employed in the device must be used as the fundamental degrees-of-freedom during the synthesis process.

In this scenario, the Proceedings of the Third Edition of the International Workshop on Metamaterials-by-Design: Theory, Methods, and Applications to Communications and Sensing, which was held in Madrid, Spain, on December 14-15, 2017, are specifically devoted to the presentation of an overview of the advances in the theory and applications of the MbD paradigm as applied to Communication and Sensing technologies, and at discussing some of the most important challenges currently open in the field.

GIACOMO OLIVERI, Editor
EVA RAJO-IGLESIAS, Editor

Acknowledgment

This work benefited from the networking activities carried out within the Càtedra de Excelencia UC3M-Santander funded by the Universidad Carlos III de Madrid (Spain) (Prof. A. Massa) and from the networking activities carried out within the FET Open CSA PROJECT NANOARCHITECTRONICS [grant Agreement number 737135] funded by the European Union.

Content from this work may be used under the terms of the Creative Commons Attribution 3.0 licence. Any further distribution of this work must maintain attribution to the author(s) and the title of the work, journal citation and DOI.
Published under licence by IOP Publishing Ltd

IOP Conf. Series: Journal of Physics: Conf. Series **963** (2018) 011002 doi:10.1088/1742-6596/963/1/011002

Peer review statement

All papers published in this volume of *Journal of Physics: Conference Series* have been peer reviewed through processes administered by the proceedings Editors. Reviews were conducted by expert referees to the professional and scientific standards expected of a proceedings journal published by IOP Publishing.

Content from this work may be used under the terms of the Creative Commons Attribution 3.0 licence. Any further distribution of this work must maintain attribution to the author(s) and the title of the work, journal citation and DOI.
Published under licence by IOP Publishing Ltd

IOP Conf. Series: Journal of Physics: Conf. Series **963** (2018) 012001 doi:10.1088/1742-6596/963/1/012001

Artificial high effective permittivity medium in a SIW filled with metallic cylinders

G. Vicent, E. Bronchalo, A. Coves and G. Torregrosa

Departamento de Ingeniería de Comunicaciones, Universidad Miguel Hernández de Elche, Spain

E-mail: ebronchalo@umh.es, angela.coves@umh.es, gtorregrosa@umh.es

Abstract. A new topology of step-impedance band-pass filters in Substrate Integrated Waveguide (SIW) technology has been recently demonstrated in which low effective permittivity regions have been achieved by removing part of the substrate material and then shielding the perforated structure. Alternatively, in this work a new way to obtain an increased relative permittivity in the guiding region is proposed by periodically inserting metallic inclusions. This paper shows the results of a systematic study of the effective permittivity obtained in this way in a SIW in order to synthesize a higher effective permittivity, which can be used in the filter design.

1. Introduction

Recently, a new topology of step-impedance bandpass filters in SIW technology has been theoretically and experimentally demonstrated [1],[2]. The high impedance regions, i.e., those with low effective permittivity, have been achieved by inserting air holes to lower the substrate permittivity. However, there is a natural limit in the low effective permittivity obtained in the perforated regions, given by the maximum volume of dielectric material that can be removed. In [1], a reduction of the relative permittivity in a perforated SIW waveguide of 42% was achieved, which was theoretically and experimentally demonstrated. Alternatively, in this work we propose a way to obtain an increased relative permittivity in the guiding region by inserting in the dielectric an array of metallic inclusions, as demonstrated in previous works [3],[4]. This paper presents the numerical results of a study of the effective permittivity in a SIW in which metallic cylinders have been periodically inserted to synthesize a higher effective permittivity. The study has been performed by using the commercial software tool Ansys High Frequency Structural Simulator (HFSS). Several insertion densities have been simulated to study the influence of the metal/dielectric volume relation in the effective permittivity. Finally, our preliminary results have been compared to the predictions of a theoretical model of this artificial propagation medium [5] to provide a helpful insight for designing new devices.

2. Study of an artificial high effective permittivity medium in a SIW filled with metallic cylinders

Fig. 1(a) shows a SIW section, with two rows of parallel metallic posts (or via holes) characterized by their separation s_v and diameter d_v, delimiting the area of propagation of the waveguide. In our case, the value of the waveguide height is $b = 0.63$ mm, corresponding to

Content from this work may be used under the terms of the Creative Commons Attribution 3.0 licence. Any further distribution of this work must maintain attribution to the author(s) and the title of the work, journal citation and DOI.
Published under licence by IOP Publishing Ltd

IOP Conf. Series: Journal of Physics: Conf. Series **963** (2018) 012001 doi:10.1088/1742-6596/963/1/012001

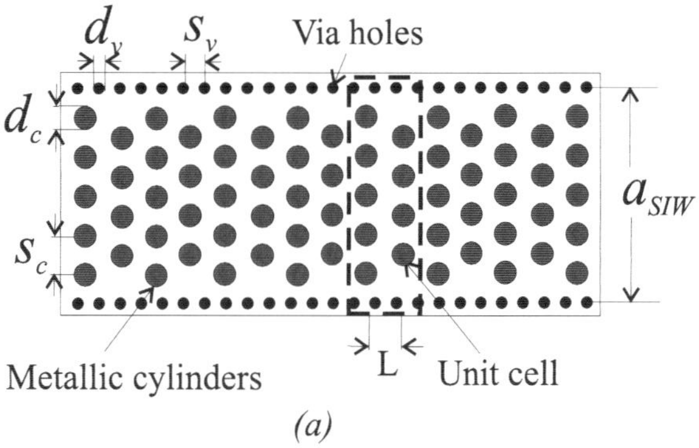

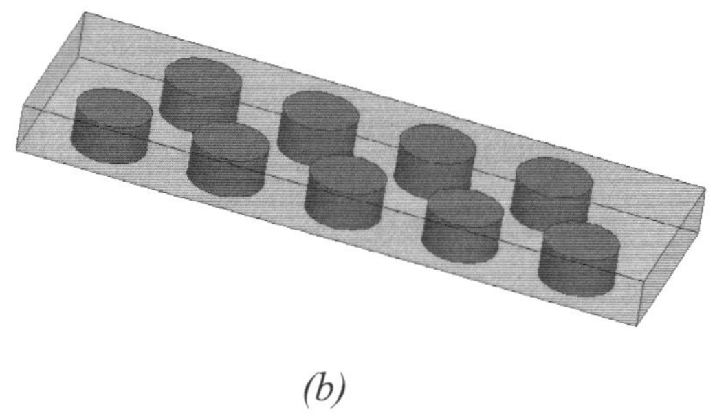

Figure 1. (a) Section of a SIW structure with periodic metallic cylinders. (b) Unit cell of the equivalent rectangular waveguide filled with periodic metallic cylinders.

the thickness of the substrate employed. In this SIW, metallic cylinders characterized by their height h, diameter d_c and separation s_c -in this particular triangular pattern configuration- have been added along the propagation region. Thus, the propagation constant of this guide is determined by the width a_{SIW} (see Fig. 1(a)) of the SIW, and also by the effective permittivity obtained after the insertion of the metallic cylinders, whose height h must be lower than, but not far from, the waveguide height b to achieve a high effective permittivity in the waveguide.

Using the equivalence between a SIW and a rectangular waveguide of a slightly lower width a [6], we have characterized the electromagnetic field propagation in this equivalent waveguide with periodic metallic cylinders. In our study, we have obtained the dispersion curve of the waveguide modes with the commercial software tool Ansys HFSS, which also yields the cutoff frequency of the modes. For designing purposes, we can restrict our study to the monomode regime of the waveguide. In this case, it is interesting to define an equivalent dielectric homogeneous waveguide whose effective permittivity is related to the cutoff frequency of its TE_{10} mode through the following expression:

$$\varepsilon_{ref} = \frac{c^2}{4a_{eff}^2 f_c^2} \tag{1}$$

where c is the speed of light in free space and f_c is the cutoff frequency of the first waveguide mode provided by the simulation software.

Using eq. (1), we have obtained the effective relative permittivity of a rectangular waveguide of dimensions $a = 8.5$ mm and $b = 0.63$ μm (being b the height of the substrate employed –Taconic CER-10–, with $\varepsilon_r = 10$ and $\tan\delta = 0.0035$), periodically filled with a triangular pattern of metallic cylinders along the waveguide, with alternating columns of 5 and 4 cylinders of height $h = 4/5\ b$ along the waveguide for this particular configuration (see Fig. 1(b)).

Firstly, we have obtained the effective permittivity of this partially filled rectangular waveguide with metallic cylinders as a function of their diameter d_c, with a fixed distance between their centers of $s_c = 1.6$ mm. In Fig. 2 it is represented the effective permittivity of this waveguide as a function of d_c, where it can be seen that $\varepsilon_{\mathrm{ref}}$ varies from $\varepsilon_{ref} = 10$ (when $d_c = 0$, i.e., for the case without metallic cylinders) up to 22.8 (for $d_c = 1.1$ mm). This graph can be employed in order to synthesize a desired ε_{ref} for a filter design.

Next, we have analyzed the variation of the effective permittivity with the distance between cylinders, for a fixed diameter $d_c = 1.1$ mm. In Fig. 3 it is represented the effective permittivity of this waveguide as a function of s_c from 1.3 mm (corresponding to a minimum distance between the edges of adjacent cylinders of 200 μm, in order to ensure a certain mechanical strength) to

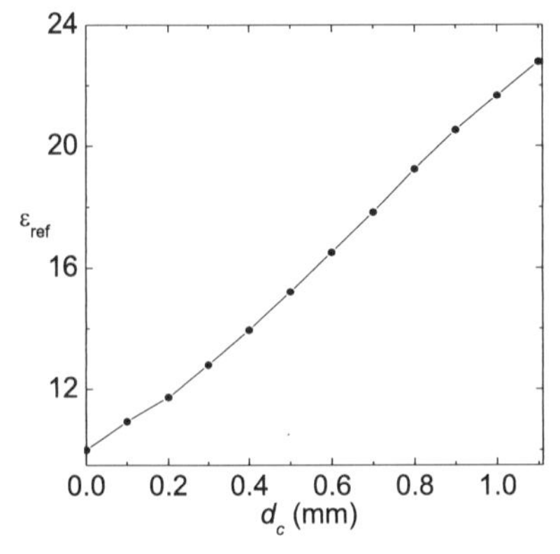

Figure 2. ε_{ref} as a function of d_c, for a fixed separation $s_c = 1.6$ mm.

1.7 mm (distance between the edges of 600 μm). Again, this graph allows to synthesize a desired ε_{ref} for designing purposes.

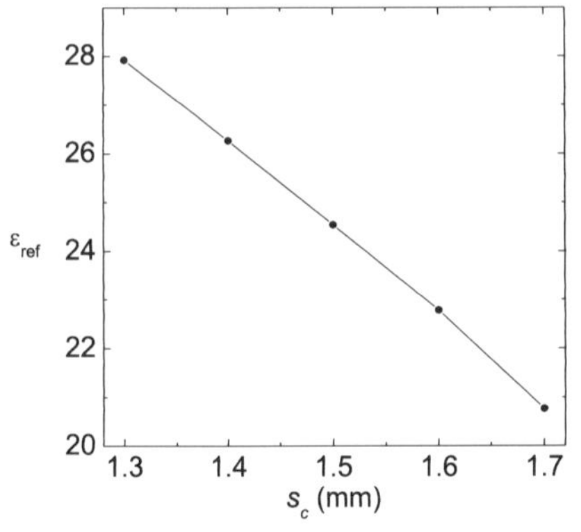

Figure 3. ε_{ref} as a function of s_c, for a fixed diameter $d_c = 1.1$ mm.

At the sight of the previous results, we have chosen an implementable configuration of $d_c = 1.1$ mm and $s_c = 1.6$ mm (with a distance between the edges of 0.5 mm), which yields an effective permittivity of $\varepsilon_{ref} = 22.8$. For this configuration, we have computed the dispersion diagram of the periodically filled waveguide with metallic cylinders. In Fig. 4 it is compared the dispersion curve of the periodically filled waveguide with metallic cylinders for $d_c = 1.1$ mm and $s_c = 1.6$ mm (crosses) with that of its equivalent dielectric homogeneous waveguide (solid line) of $\varepsilon_{ref} = 22.8$. In this figure, it can be checked that the effective permittivity equivalence is only strictly satisfied at cutoff, given the frequency dependent nature of the metallic inclusions. However, the equivalence can be used as a first approximation in the filter design.

In order to check the validity of the previous results, we have analyzed the dispersion behaviour of a periodically filled SIW with metallic cylinders. To this end, we have obtained the equivalent rectangular width in SIW technology a_{SIW}, with the following parameters for the via holes: $d_v = 0.7$ mm, $s_v = 0.95$ mm, which provide a width of SIW of $a_{SIW} = 9.2$ mm. On the other hand, we have designed microstrip to SIW transitions consisting of microstrip tapers [7] (see Fig. 5), whose dimensions W_t and L_t have been optimized for a maximum matching of the waveguide transitions, while the width of the microstrip line is 0.6 mm (50 Ω). In Fig. 6 there are shown the scattering parameters of a 46 mm length section of the periodically filled SIW with metallic cylinders under study (including dielectric and conductor losses -copper-) obtained with Ansys HFSS, showing a cutoff frequency of $f_c = 3.68$ GHz, which is very similar to the value previously obtained of 3.70 GHz in the dispersion curve with Ansys HFSS for the same waveguide configuration.

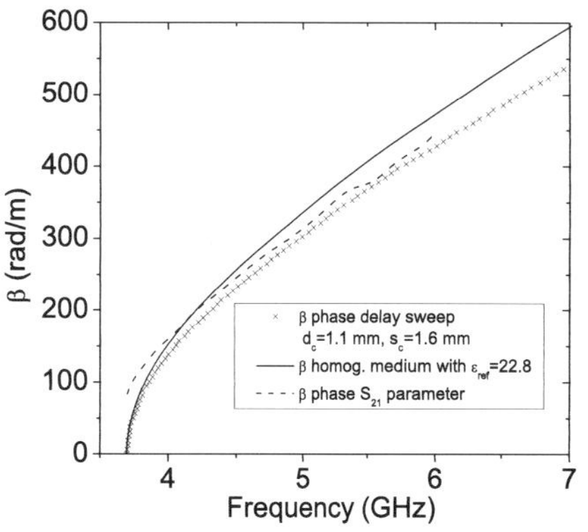

Figure 4. Dispersion curve of the periodically filled waveguide with metallic cylinders for $d_c = 1.1$ mm and $s_c = 1.6$ mm.

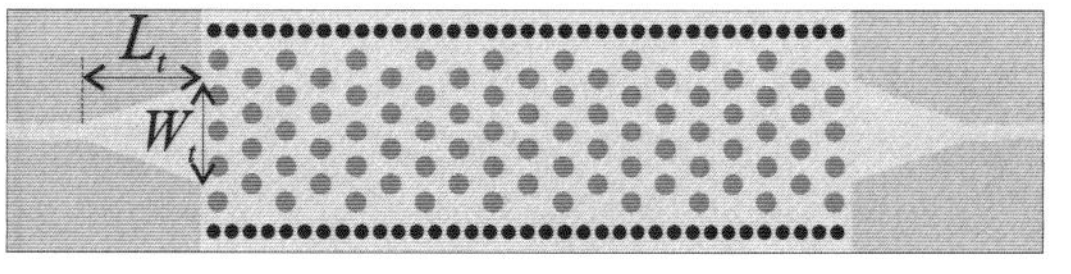

Figure 5. Scheme of a periodically filled SIW with metallic cylinders.

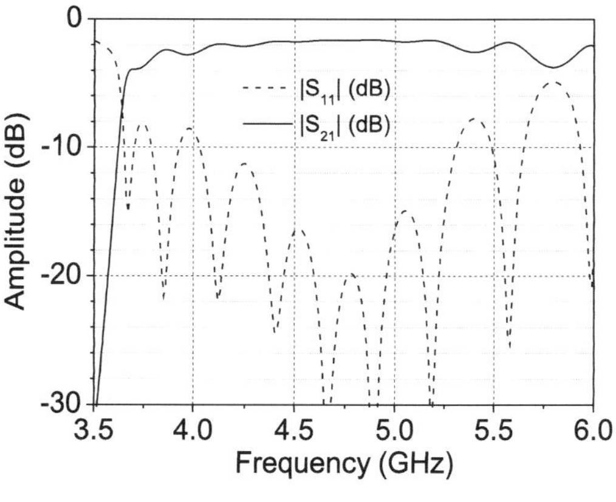

Figure 6. Electrical response of a SIW section of width $a = 9.2$ mm and length $l = 46$ mm periodically filled with metallic cylinders of parameters: $d_c = 1.1$ mm and $s_c = 1.6$ mm.

Finally, the dispersion curve of the SIW waveguide periodically filled with metallic cylinders whose electrical response is shown in Fig. 6 has been computed from its scattering parameters obtained for two different lengths using the same approach as that presented in [8]. The simulated dispersion curve obtained by this method has been presented in Fig. 4 with dashed line, showing a good agreement in the represented operation band. All the theoretical analysis performed in this work is going to be experimentally verified shortly.

3. Theoretical modeling and comparison with simulations

A theoretical model of the artificial propagation medium is convenient to provide a better understanding of the electromagnetic problem and helpful insight for designing new devices. The theoretical study of artificial dielectrics began early in the history of electromagnetism [5]. Some particular problems have been solved in the context of electrostatics, as the one that we have chosen to model the propagation in the cylinder-loaded SIW: a tetragonal array of conducting cylinders oriented along one of the array axis and with the external electric field parallel to the cylinders. The effective permittivity relative to the host medium permittivity $\widetilde{\varepsilon}_{ref}$ is given by [9]

$$\widetilde{\varepsilon}_{ref} = 1 + \frac{N\alpha/\varepsilon_0}{1 - C\alpha/\varepsilon_0}. \tag{2}$$

where α is the polarizability of the inclusions, N the number of inclusions per unit volume, and C is an interaction constant providing the correction to the external electric field due to the dipoles surrounding the dipole under consideration. For an infinite tetragonal network of dipoles, the interaction constant is given by [9]

$$C = \frac{1}{l_y^3}\left\{\frac{1.202}{\pi} - 16\pi K_0\left(\frac{2\pi l_t}{l_y}\right)\right\} \tag{3}$$

In this expression, Y is the axis parallel to the external electric field, l_t are the unit cell dimensions along the X and Z axes, while l_y is the unit cell dimension along the Y axis, and K_0 is the modified Bessel function of second kind and zero order.

Applying the image theory to the waveguide walls allows us to represent the waveguide as an infinite dipole array. We will place just a layer of cylindrical insertions parallel to the wider side of the waveguide. Therefore, $l_y = b$ in eq. (3), being b the height of the waveguide.

Referring the polarizability of a perfectly conducting cylinder, we have followed the paper of Taylor [10]. The polarizability for the field oriented along the cylinder axis is given by

$$\frac{\alpha}{\varepsilon_0 V_0} = \frac{2h}{D}\left[\frac{r_b}{3} + r_0\frac{\Gamma(3/2)}{2^{2/3}\Gamma(13/6)}\right] + t_b + t_0\frac{2^{1/3}}{\Gamma(5/3)} \tag{4}$$

where h is the height of the cylinder, D its diameter and V_0 its volume, and where the parameters r_b, r_0, t_b and t_0 depend on the aspect ratio h/D [10]. For the substrate presented in the preceding sections, with $b = 0.63$ mm, we have chosen as feasible dimensions $h = D = 0.5$ mm. We have simulated with Ansys HFSS a waveguide with $a = 10$ mm and several densities of inclusions: 4 to 9 cylinders along the wider side of the waveguide, corresponding to l_t between 1.1 and 2.5 mm. We have left a distance $l_t/2$ from the outermost cylinders to the vertical walls of the waveguide to get the same spacing up to the first image dipole formed by these walls. The effective permittivity of this transmission media has been obtained from the simulation by the method explained in the last section.

The results are compared to the predictions of the theoretical model expressed by formulas 2-4 in Fig. 7. It is clear that the model predictions scale much faster with the density of inclusions. This divergence could be due to several causes: 1) The dipole approximation may be inadequate for high inclusion densities; 2) The inhomogeneity of the electric field for the guided mode is not considered in the model; 3) The theoretical model is electrostatic, while the propagating mode is electromagnetic. Those aspects of the problem are now under consideration to elaborate a more sophisticated model for the SIW effective permittivity.

4. Conclusion

In this work it is proposed a new way to increase the permittivity in a SIW by inserting a periodic array of metallic cylinders along the waveguide. A systematic study of the resulting effective permittivity has been performed as a function of the cylinders diameter and separation. The results of this study have been numerically checked through simulations of the electrical response of a particular design of a SIW filled with metallic cylinders.

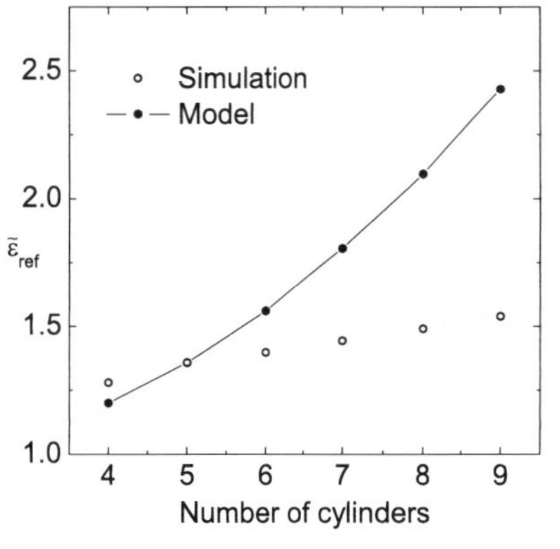

Figure 7. $\tilde{\varepsilon}_{ref}$ of a waveguide loaded with conducting cylinders: results of Ansys HFSS simulation and theoretical model.

Acknowledgments
This work was supported by the Agencia Estatal de Investigación (AEI) and by the Unión Europea through the Fondo Europeo de Desarrollo Regional - FEDER - "Una manera de hacer Europa" (AEI/FEDER, UE), under the Research Project TEC2016-75934-C4-2-R.

References
[1] Coves A, Torregrosa G, San Blas A A, Sánchez-Soriano M A, Martellosio A, Bronchalo E and Bozzi M 2016 *Radio Science* **65** 328–36
[2] Silvestri L, Massoni E, Tomassoni C, Coves A, Bozzi M and Perregrini L 2017 *IEEE Trans. Microw. Theory Tech.* **51** 2687–97
[3] Nguyen H V, Gauthier J, Fernández J M, Sierra-Castañer M and Caloz C *Proc. of Asia-Pacific Microwave Conference* 2006 Yokohama Japan
[4] Njoku C C, Whittow W G and Vardaxoglou J C 2012 *IEEE Trans. Antennas and Prop.* **60** 2194–204
[5] Kharadly M M Z and Jackson W 1953 *Proc. of the IEE - Part III: Radio and Communic. Engin.* **100** 199–212
[6] Cassivi Y, Perregrini L, Arcioni P, Bressan M, Wu K and Conciauro G 2002 *IEEE Microw. Wirel. Compon. Lett.* **12** 333–35
[7] Deslandes D and Wu K 2001 *IEEE Microw. Wireless Comp. Lett.* **11** 68–70
[8] Deslandes D, Bozzi M, Arcione P and Wu K 2003 *Proc. of the IEEE MTT-S Int. Microw. Symp. Dig.* **3** 1103–6
[9] Collin R E 1990 *Field theory of guided waves* (Wiley-IEEE Press)
[10] Taylor T T 1960 *J. Research of the National Bureau of Standards- B. Mathematics and Mathem. Phys.* **64B** 135–43

Exploiting Glide Symmetry in Planar EBG Structures

Boules A. Mouris, Oscar Quevedo-Teruel and Ragnar Thobaben
School of Electrical Engineering, KTH Royal Institute of Technology, Stockholm, Sweden.
E-mail: boules@kth.se

Abstract. Periodic structures such as electromagnetic band gap (EBG) structures can be used to prevent the propagation of electromagnetic waves within a certain frequency range known as the stop band. One of the main limitations of using EBG structures at low frequencies is their relatively large size. In this paper, we investigate the possibility of using glide symmetry in planar EBG structures to reduce their size. Simulated results demonstrate that exploiting glide symmetry in EBG structures can lead to size reduction.

1. Introduction

Electromagnetic band gap (EBG) structures are well-known for controlling the propagation of electromagnetic waves in a certain frequency band. Via holes, or short-circuits, are conventionally employed to produce these EBGs [1]. However, these vias increase the cost of manufacturing. To overcome this drawback, a completely planar EBG structure made of metallic patches only without via holes was proposed in [2] for mutual coupling reduction in patch antenna arrays.

Higher symmetries in periodic structures was first introduced in the 70's [3]. Glide symmetry is a type of higher symmetries in which a unit cell is translated half its period and mirrored in a glide plane [4–7]. The interesting properties of glide-symmetric structures have been recently investigated for a number of applications such as: gap-waveguide technology [8, 9], ultra wide band lens antennas [10, 11]. Moreover, the recent discoveries in the field of glide symmetries has promoted new research in other higher symmetries, as twist symmetry [12].

In this paper, a planar EBG structure with glide symmetry is proposed and compared to the case with no glide symmetry.

2. Proposed structure and simulated results

The unit cell of the proposed glide-symmetric EBG structure is shown in Fig. 1. The structure is made of square metallic patches similar to those used in [2] for comparison purposes. Two layers of a high permittivity material ($\epsilon_r = 10$) of thickness $t = 1.27$ mm are assumed. The dimensions of the unit cell were initially selected to be $d_x = d_y = 25$ mm and $p_x = p_y = 15$ mm. For the case of no glide symmetry, the square patches are placed on the upper layer with no patches on the first layer as shown in Fig. 2 in order to keep the thickness of the substrate constant in both cases.

Figure 3a shows the dispersion diagram of the EBG structure in the x direction in both cases with and without glide symmetry. These dispersion diagrams have been calculated with the

Content from this work may be used under the terms of the Creative Commons Attribution 3.0 licence. Any further distribution of this work must maintain attribution to the author(s) and the title of the work, journal citation and DOI.
Published under licence by IOP Publishing Ltd

eigenmode analysis of commercial software. These dispersion diagrams shows that in the case of glide symmetry, the first mode is degenerated into two modes which are connected in Brillouin zone. This is a characteristic of higher-symmetric structures [7]. A finite structure of size 5×5 unit cells was then simulated with a transient solver, including in this case two physical ports. The results for the transmission coefficient S_{21} are shown in Fig. 3b. These simulation results show that the stop band of the glide-symmetric structure has a lower frequency appearance than the one with no glide symmetry for the same dimensions. Meaning that, a more compact EBG structure can be used with glide symmetry to obtain the same stop band.

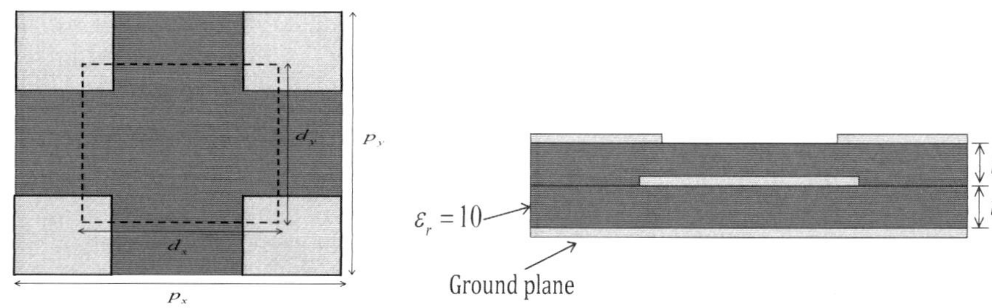

(a) Top view (b) Front view

Figure 1: Unit cell of the glide-symmetric periodic structure

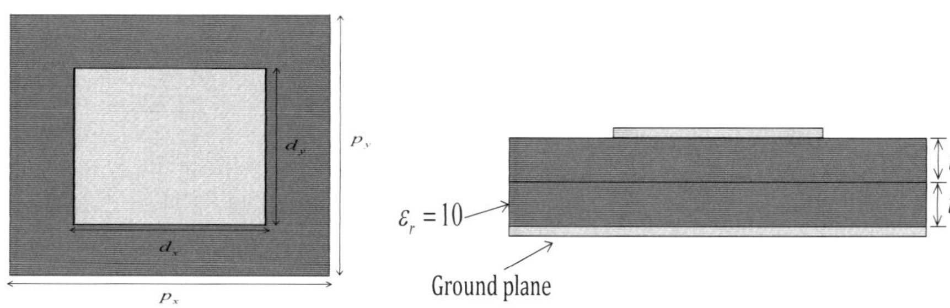

(a) Top view (b) Front view

Figure 2: Unit cell of the periodic structure with no glide symmetry

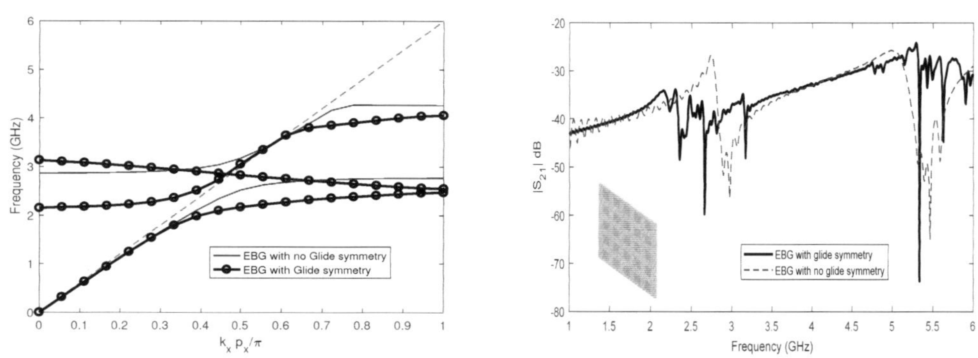

(a) Dispersion diagram of the EBGs (b) Transmission coefficient (S_{21})for a 5 unit cells finite structure

Figure 3: Simulation results obtained for EBGs made of patches with and without glide symmetry.

IOP Conf. Series: Journal of Physics: Conf. Series **963** (2018) 012002 doi:10.1088/1742-6596/963/1/012002

3. Conclusions

A planar EBG structure with glide symmetry was proposed in this paper and compared to the case with no glide symmetry. The dispersion diagram as well as the transmission coefficient for a finite structure were presented showing that glide symmetry can be applied to planar EBG structures to miniaturize their size.

References

[1] Sievenpiper D, Zhang L, Broas R F J, Alexopolous N G and Yablonovitch E 1999 *IEEE Transactions on Microwave Theory and Techniques* **47** 2059–2074 ISSN 0018-9480

[2] Rajo-Iglesias E, Quevedo-Teruel O and Inclan-Sanchez L 2008 *IEEE Transactions on Antennas and Propagation* **56** 1648–1655 ISSN 0018-926X

[3] Hessel A, Chen M H, Li R C M and Oliner A A 1973 *Proceedings of the IEEE* **61** 183–195 ISSN 0018-9219

[4] Crepeau P J and McIsaac P R 1964 *Proceedings of the IEEE* **52** 33–43 ISSN 0018-9219

[5] Kieburtz R and Impagliazzo J 1970 *IEEE Transactions on Antennas and Propagation* **18** 3–7 ISSN 0018-926X

[6] Mittra R and Laxpati S 1965 *Canadian Journal of Physics* **43** 353–372 (*Preprint* `https://doi.org/10.1139/p65-032`) URL `https://doi.org/10.1139/p65-032`

[7] Valerio G, Sipus Z, Grbic A and Quevedo-Teruel O 2017 *IEEE Transactions on Antennas and Propagation* **65** 2695–2700 ISSN 0018-926X

[8] Ebrahimpouri M, Quevedo-Teruel O and Rajo-Iglesias E 2017 *IEEE Microwave and Wireless Components Letters* **27** 542–544 ISSN 1531-1309

[9] Ebrahimpouri M, Rajo-Iglesias E, Sipus Z and Quevedo-Teruel O 2017 *In Press IEEE Transactions on Microwave Theory and Techniques*

[10] Liu K, Ghasemifard F and Quevedo-Teruel O 2017 *11th European Conference on Antennas and Propagation (EUCAP)* pp 358–360

[11] Quevedo-Teruel O, Ebrahimpouri M and Kehn M N M 2016 *IEEE Antennas and Wireless Propagation Letters* **15** 484–487 ISSN 1536-1225

[12] Dahlberg O, Mitchell-Thomas R C and Quevedo-Teruel O 2017 *Scientific Reports* **7** 10136 ISSN 2045-2322 URL `https://doi.org/10.1038/s41598-017-10566-w`

Shape design of a reflecting surface using Bayesian Optimization

X. Garcia-Santiago[1,2], P.-I. Schneider[1], C. Rockstuhl[2,3], S. Burger[1,4]

[1] JCMwave GmbH., 12163 Berlin, Germany
[2] Institute of Nanotechnology, Karlsruhe Institute of Technology, 76021 Karlsruhe, Germany
[3] Institute of Theoretical Solid State Physics, Karlsruhe Institute of Technology, 76131 Karlsruhe, Germany
[4] Zuse Institute Berlin, 14195 Berlin, Germany

E-mail: xavi.garcia@jcmwave.com

Abstract. Optimizing free-from shapes of photonic nanostructures is a high-dimensional problem. We present a Bayesian optimization algorithm with a hyper-parameter learning routine. We apply the algorithm for optimizing the shape of a reflecting meta-surface. The algorithm exhibits efficient performance.

1. Introduction

Nanooptical technologies have drastically evolved in recent years with applications in fields such as silicon photonics, energy harvesting, or quantum technologies. This improvement is mainly due to modern fabrication technologies such as laser writing [1] or electron-beam lithography [2] which allow for the manufacturing of micro and nano optical structures with an increasing degree of accuracy and flexibility. However, the best shape of the photonic devices for achieving a good performance is not always easy to find. The most efficient designs can be sometimes counter-intuitive. In these cases inverse engineering methods become an indispensable tool and the choice of the algorithm used to derive the shape of the device becomes a critical aspect.

One important aspect of the design process is the parametrization used in order to describe the shape of the device, this parametrization should be flexible enough in order to be able to describe a high number of possible shapes. However, this flexibility presents two main drawbacks. First, it normally implies working in high dimensional parameter spaces, which directly increases the computational time needed to find the optimal point. Second, the more general the shape is allowed to be, the more likely it is that the optimization problem is not convex, leading to problems where the objective function presents many local minima.

Many algorithms have been used in order to globally optimize different nanophotonic structures. From genetic [3] or evolution [4] algorithms, direct-binary search (DBS) [5], particle swarm [6] or more recently via the use of machine learning algorithms [7]. Into this last group we could include Bayesian optimization. An algorithm based on a stochastic model where the fundamental parameters of the algorithm are updated during computation time based on the information acquired from the previous simulations. This group has recently presented a study on the use and performance of this algorithm in order to optimize different optical nano-structures [8]. We now update this work presenting an improvement in the learning process.

Content from this work may be used under the terms of the Creative Commons Attribution 3.0 licence. Any further distribution of this work must maintain attribution to the author(s) and the title of the work, journal citation and DOI.
Published under licence by IOP Publishing Ltd

This improvement leads to better convergence ratios, thereby allowing the access to parameter spaces of higher dimensions and therefore, to more complex structures. As an example, we optimize the shape of a periodic meta-mirror in order to increase the power reflected into the first diffraction order.

2. Bayesian Optimization with Gaussian Processes

The goal of every global optimization algorithm is to identify the global minimum (or maximum) of an unknown objective function f in a certain design space $\mathcal{X}$,

$$\mathbf{x_{min}} = \arg\min_{\mathbf{x}\in\mathcal{X}} f(\mathbf{x}), \tag{1}$$

In order to achieve that, Bayesian optimization (BO) uses a stochastic model of the objective function f. The model is updated with information from the evaluations of f and then it is used to drive the optimization strategy.
Gaussian processes (GP) are frequently used as the stochastic model in Bayesian optimization. Our objective function is modeled as a Gaussian process if for any N points $\mathbf{x}_1, \cdots \mathbf{x}_N$ in the parameter space $\mathcal{X}$, the probability of the objective function to be equal to $\mathbf{Y} = (y_1, \cdots, y_N)$ at these points is following a multivariate Gaussian random distribution defined by a mean function $\mu : \mathcal{X} \to \mathbb{R}$ and a positive definite covariance function $k : \mathcal{X} \times \mathcal{X} \to \mathbb{R}$,

$$P(\mathbf{Y}) = \frac{1}{(2\pi)^{N/2}|\mathbf{\Sigma}|^{1/2}} \exp\left[-\frac{1}{2}(\mathbf{Y}-\mu)^T\mathbf{\Sigma}^{-1}(\mathbf{Y}-\mu)\right] \tag{2}$$

or more compactly $\mathbf{Y} \sim \mathcal{N}(\mu, \mathbf{\Sigma})$. Here, $\mu = [\mu(\mathbf{x}_1), \cdots, \mu(\mathbf{x}_k)]^T$ is the vector of mean values and $\mathbf{\Sigma} = [k(x_i, x_j)]_{i,j}$ the covariance matrix.

GP can describe a large class of random functions using different covariance functions $k(\mathbf{x}, \mathbf{x}')$. For our optimization purposes and based on the results obtained in [8], we will use the Matérn 5/2 kernel,

$$k_{\mathrm{M52}}(\mathbf{x}, \mathbf{x}') = \sigma^2\left(1 + \sqrt{5r^2(\mathbf{x}, \mathbf{x}')} + \frac{5}{3}r^2(\mathbf{x}, \mathbf{x}')\right)\exp\left(-\sqrt{5r^2(\mathbf{x}, \mathbf{x}')}\right) \tag{3}$$

$$r^2(\mathbf{x}, \mathbf{x}') = \sum_{i=1}^{d}(x_i - x_i')^2/l_i^2 \tag{4}$$

The hyper-parameters σ and $l_1, l_2, \cdots$ determine the standard deviation and the length scales of the random distributions. With respect to the mean vector μ, we will assume all its elements have the same constant value m. The different choices for μ do not influence the performance of the method as much as the kernel function does.

The stochastic model is then updated after the evaluations of the objective function using Bayes' theorem. In the points where the problem has been solved, the value of the objective is no longer an unknown but rather a well known value without any uncertainty. Also in the regions close to the evaluated points the uncertainty will be reduced. The posterior probability distribution is computed based on the computed objective evaluations and on the prior distribution of the geometry parameters.

Suppose M values $\mathbf{Y_{ev}} = (y_{ev1}, \cdots y_{evM}) = (f(\mathbf{x_{ev}}_1), \cdots, f(\mathbf{x_{ev}}_M))$ of the unknown objective function are known. Then the posterior distribution will also be a Gaussian process G'. The new mean vector μ' and the covariance matrix $\mathbf{\Sigma}'$ are functions of the previous mean vector μ, of the covariance matrix $\mathbf{\Sigma}$ and also of its values in the evaluation points $\mathbf{x_{ev}}_i$

$$\mu'(\mathbf{x}) = \mu(\mathbf{x})^T + \mathbf{\Sigma}_{12}\mathbf{\Sigma}_{22}^{-1}(\mathbf{Y_{ev}} - \mu(\mathbf{x_{ev}})^T) \tag{5}$$

$$\mathbf{\Sigma}' = \mathbf{\Sigma}_{11} - \mathbf{\Sigma}_{12}\mathbf{\Sigma}_{22}^{-1}\mathbf{\Sigma}_{21}) \tag{6}$$

with $(\mathbf{\Sigma}_{11})_{i,j} = k(\mathbf{x}_i, \mathbf{x}_j)$, $(\mathbf{\Sigma}_{12})_{i,j} = k(\mathbf{x_{ev}}_i, \mathbf{x}_j)$, and $(\mathbf{\Sigma}_{22})_{i,j} = k(\mathbf{x_{ev}}_i, \mathbf{x_{ev}}_j)$.

In Fig. 1 we show a stochastic model of an objective function before (left) and after (right) being updated with five evaluations.

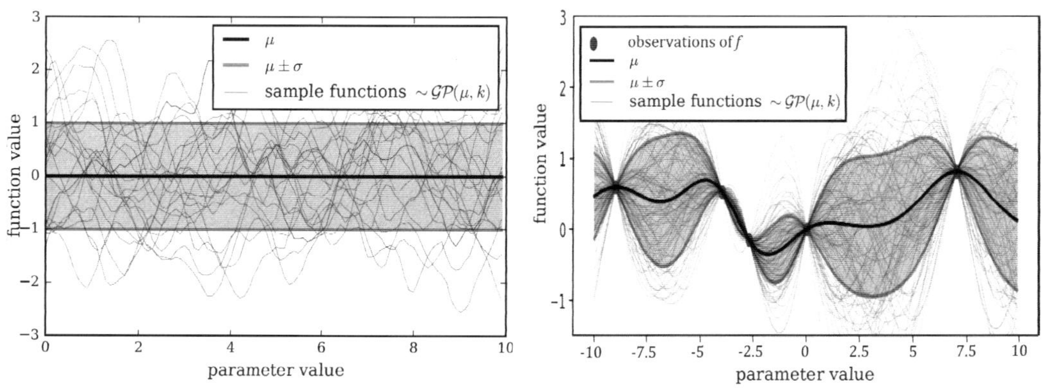

Figure 1. a. Different samples of a Gaussian process with the Matérn 5/2 kernel defined in Eq. (3) **b**. Different samples of a Gaussian process with the Matérn 5/2 kernel updated after five evaluations of the sample objective function $f(x) = 0.5\sin(x) + 0.01x^2$.

Finally, the optimization strategy i.e., the decision where to make the next evaluation is determined by an *acquisition function*. In our case we use the expected improvement function α_{EI},

$$\begin{aligned}\alpha_{\mathrm{EI}}(x, y_{\min}) &= \mathbb{E}[\max(0, y_{\min} - f(x))] \\ &= \frac{1}{2}\left[1 + \mathrm{erf}\left(\frac{\mathrm{y}_{\min} - \mu_{\mathrm{N}}(\mathbf{x})}{\sqrt{2}\sigma_{\mathrm{N}}(\mathbf{x})}\right)\right](y_{\min} - \mu_N(\mathbf{x})) + \frac{\sigma_N(\mathbf{x})}{\sqrt{2\pi}}\exp\left(\frac{(y_{\min} - \mu_N(\mathbf{x}))^2}{2\sigma_N(\mathbf{x})^2}\right)\end{aligned} \tag{7}$$

In Fig. 2 an example of the expected improvement function is shown. With this function, the algorithm selects the next point to evaluate based on where the stochastic model expects the largest reduction in the currently computed minimum.

3. The hyper-parameters learning process

Initially, the hyper-parameters of the GP $\omega = (\sigma, l_1, l_2, \cdots)$ are unknown and are usually initialized with some reasonable values. Having drawn some samples of the objective function one can optimize the hyper-parameters by maximizing the probability of drawing these samples with respect to the values of the hyper-parameters:

$$\omega_{\mathrm{opt}} = \arg\max_{\omega}\left(\log[P_\omega(\mathbf{Y_{ev}})] + \log[P(\omega)]\right) \tag{8}$$

Here, $P(\omega)$ is the prior distribution of the hyper-parameters (the hyper-prior) and $P_\omega(\mathbf{Y_{ev}})$ the likelihood of observing $\mathbf{Y_{ev}}$ given ω.

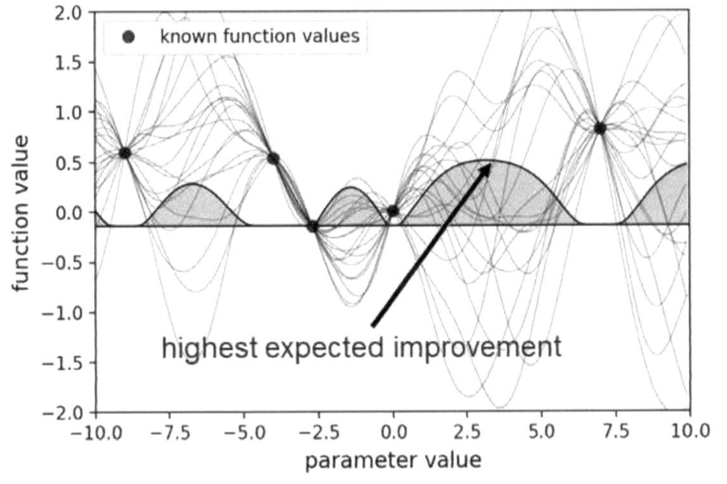

Figure 2. The optimization strategy evaluates the objective function in the point with the highest expected improvement (eq. 7)

The value of the hyper-parameters is crucial for an efficient optimization. A too small length scale, for example, would overestimate the uncertainty in points close to the evaluated points. However, if it is too large, the acquisition function would tend to point to the boundaries of the parameter space and to under-sample the central regions. However, every time the hyper-parameters are updated the covariance matrix has to be inverted, an expensive task which slows down the optimization algorithm considerably, specially for high-dimensional parameter spaces. One solution to overcome this problem is to not update the hyper-parameters after each evaluation but rather to use some measure for estimating the possible improvement (I). Then the hyper-parameters would be updated if and only if they are expected to improve the likelihood $P_\omega(\mathbf{Y_{ev}})$.

However, not all the hyper-parameters force the inversion of the covariance matrix. This is the case for μ and σ where, furthermore, analytical expressions for their optimal values can be obtained. Therefore, we can optimize these hyper-parameters after every evaluation and just leave the length scales l_i to be updated based on the possible improvement estimator I.

The log likelyhood can be written as

$$\log[P_\omega(\mathbf{Y_{ev}})] = -\frac{N}{2}log(2\pi) - \frac{1}{2}\log(|\mathbf{\Sigma}|) - \frac{1}{2}(\mathbf{Y_{ev}} - \boldsymbol{\mu})^T\mathbf{\Sigma}^{-1}(\mathbf{Y_{ev}} - \boldsymbol{\mu}). \tag{9}$$

If we define a matrix $\mathbf{K}$ such that $\mathbf{\Sigma} = \sigma^2\mathbf{K}$, the optimal value of the hyperparameter σ can be derived analytically from

$$\frac{\partial \log[P_\omega(\mathbf{Y_{ev}})]}{\partial\sigma} = -\frac{N}{\sigma} + \frac{1}{\sigma^3}(\mathbf{Y_{ev}} - \boldsymbol{\mu})^T\mathbf{K}^{-1}(\mathbf{Y_{ev}} - \boldsymbol{\mu}) = 0. \tag{10}$$

Hence the optimal value of σ is given by the Mahalanobis distance divided by $\sqrt{N}$

$$\sigma^2_{\text{opt}} = \frac{1}{N}(\mathbf{Y_{ev}} - \boldsymbol{\mu})^T\mathbf{K}^{-1}(\mathbf{Y_{ev}} - \boldsymbol{\mu}). \tag{11}$$

Repeating the procedure for the optimal mean value m we obtain,

$$m_{\text{opt}} = \frac{\sum_{ij}(\mathbf{K}^{-1})_{ij}Y_{evj}}{\sum_{ij}(\mathbf{K}^{-1})_{ij}} \tag{12}$$

The optimization of these hyper-parameters after every evaluation is a fast calculation and it improves the convergence rate of the algorithm. Higher convergence rates allow us to investigate more complex structures as geometry parametrization spaces with higher dimensions become feasible to solve.

For the length scales l_i we use a first order approximation of the log likelihood as measure of the possible improvement,

$$I = \max_i \left(l_i \frac{\partial \log[P_\omega(\mathbf{Y_{ev}})]}{\partial l_i}\right), \tag{13}$$

The optimization of the l_i's is done provided

$$I > \epsilon \log[P_\omega(\mathbf{Y_{ev}})], \tag{14}$$

with $\epsilon = 0.1$.

4. Numerical results

In order to demonstrate the performance of BO using GP with optimized hyper-parameters we revisit the example of optimization of a reflective surface [8]. The structure consists on a periodic corrugated silver surface as shown in Fig. 3. The mirror has a periodic length of 1350 nm, 3 times the wavelength of the incident plane wave ($\lambda = 450$ nm). The optimization problem consists on finding the mirror shape which increases the power flux reflected into the first diffraction order when the mirror is illuminated with a normal incident TE polarized plane wave.

$$f_{ob} = \frac{S_{z,+1}}{S_{z,inc}} \tag{15}$$

where $S_{z,inc}$ is the z-component of the Poynting vector of the incident wave and $S_{z,+1}$ of the first diffraction order.

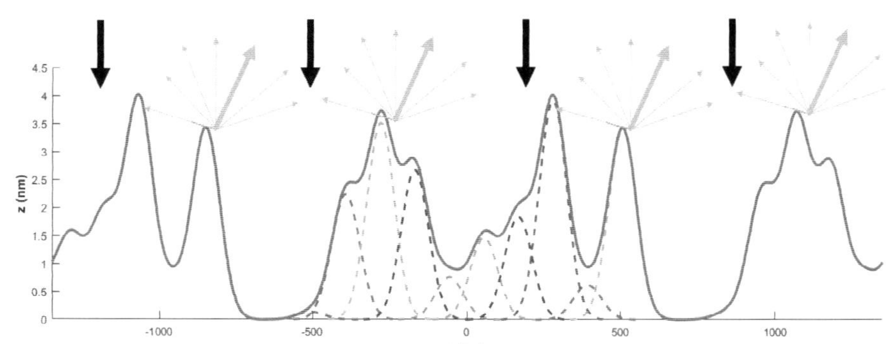

Figure 3. Schematic of the system. A plane wave illuminates the periodic mirror meta-surface (solid line) under normal incidence, which produces several diffraction orders. The objective is to find the shape of the mirror that maximizes the reflectance into the first diffraction order. The shape of the mirror is described with a parametrization of its periodic unit cell using 10 third order b-splines (dotted lines).

The mirror is parametrized with 3rd order b-splines with uniformly distributed knots as shown in Fig. 3.

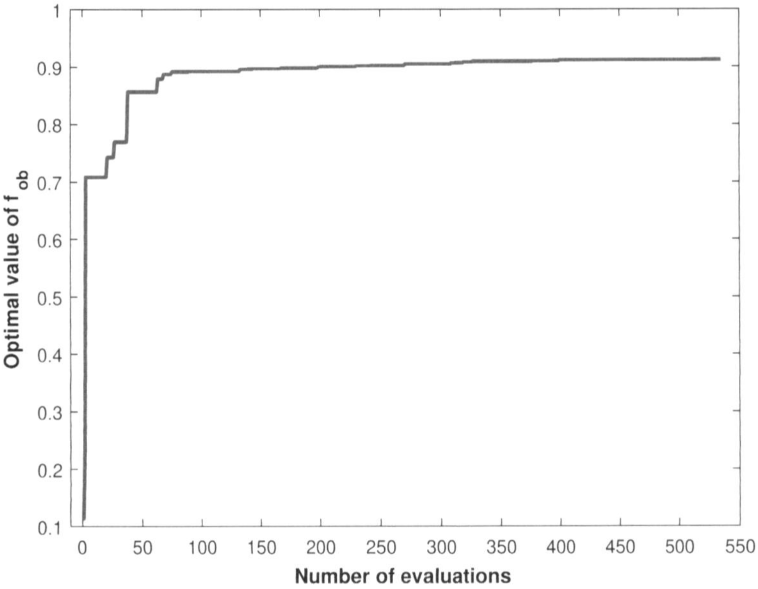

Figure 4. Convergence plot of the optimization.

The simulations of the electromagnetic field were done with the FEM solver JCMsuite [9]. We use elements with a polynomial degree of 3 and a mesh size of $\lambda/(10n(\mathbf{r}))$, where $n(\mathbf{r})$ is the refractive index of the material. The convergence results are shown in Fig. 4. The best shape deviates 91.3% of the incident power flux into the first diffraction order. The optimum shape is shown in Fig. 5. After 80 evaluations the algorithm had already found a shape with a performance of 89% and of 91% after 330 evaluations. Note that these 530 evaluations correspond to less than two points per dimension in the case of a uniformly distributed parameter scan.

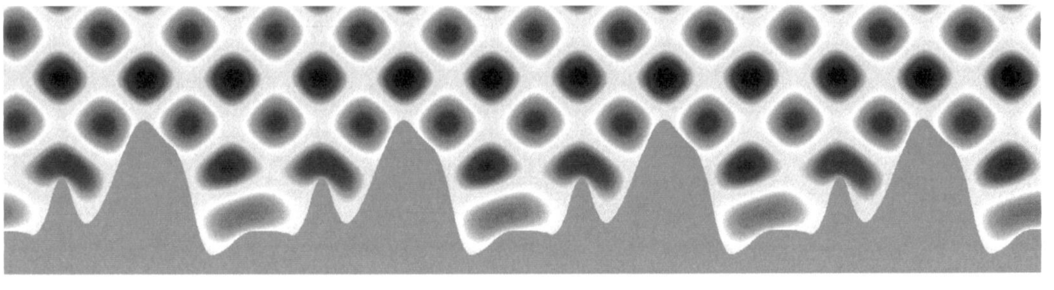

Figure 5. Electric field profile created by the optimal mirror.

5. Conclusions

We have implemented a new hyper-parameter learning routine. With the calculation of the optimum hyper-parameters σ and μ after every evaluation we get an improvement in the convergence ratio. This improvement allows us to look for shapes in higher dimensional parameter spaces, which potentially leads to a design with a better performance. We have improved the performance of a meta-surface mirror with regards to previously reported results [8] by a factor of around 25 %. This improvement is obtained thanks to the increment of the dimensions of the parameter space, going from 4 to 10 dimensions. The number of equivalent points-per-dimension needed to obtain the optimum shape was less than two. This suggests the enormous potential of BO for globally optimizing nanophotonic structures.

Acknowledgments

This project has received funding from the European Union's Horizon 2020 research and innovation programme under the Marie Skłodowska-Curie grant agreement No 675745 (MSCA-ITN-EID NOLOSS). This project has received funding from the Senate of Berlin (IBB, ProFIT grant agreement No 10160385, FI-SEQUR), co-financed by the European Fund for Regional Development (EFRE).

References

[1] Fischer J and Wegener M 2013 *Laser Photon. Rev.* **7** 22–44
[2] Chen Y 2015 *Microelectron. Eng.* **135** 57 – 72
[3] Goh J, Fushman I, Englund D and Vučković J 2007 *Opt. Express* **15** 8218–8230
[4] Preble S, Lipson M and Lipson H 2005 *Applied Physics Letters* **86** 061111
[5] Shen B, Wang P, Polson R and Menon R 2015 *Nature Photonics* **9** 378 – 382
[6] Forestiere C, Donelli M, Walsh G F, Zeni E, Miano G and Negro L D 2010 *Opt. Lett.* **35** 133–135
[7] Malkiel I, Nagler A, Mrejen M, Arieli U, Wolf L and Suchowski H 2017 *ArXiv e-prints* (*Preprint* 1702.07949)
[8] Schneider P I, Garcia-Santiago X, Rockstuhl C and Burger S 2017 *Proc. SPIE* **10335** 103350O
[9] Pomplun J, Burger S, Zschiedrich L and Schmidt F 2007 *physica status solidi (b)* **244** 3419–3434

IOP Conf. Series: Journal of Physics: Conf. Series **963** (2018) 012004 doi:10.1088/1742-6596/963/1/012004

Chiral scatterers designed by Bayesian optimization

Philipp Gutsche[1,2], Philipp-Immanuel Schneider[3], Sven Burger[2,3] and Manuel Nieto-Vesperinas[4]

[1]Freie Universität Berlin, Mathematics Institute, Arnimallee 6, 14195 Berlin, Germany
[2]Zuse Institute Berlin, Computational Nano Optics, Takustr. 7, 14195 Berlin, Germany
[3]JCMwave GmbH, Bolivarallee 22, 14050 Berlin, Germany
[4]Instituto de Ciencia de Materiales de Madrid, Consejo Superior de Investigaciones Científicas, Campus de Cantoblanco, Madrid 28049, Spain

E-mail: gutsche@zib.de

Abstract. The helicity or chirality of scattered light is strongly linked to the dual symmetry of the scatterer. The latter depends on chiral materials or on scatterers which are not superimposable with their mirror image. This inherently yields asymmetric structures of various shapes with many degrees of freedom. In order to explore these high dimensional parameter spaces, numerical simulations and especially optimization strategies are a valueable tool. Here, we optimize the emission of chiral line sources in two-dimensional dimer setups using Bayesian optimization. We deduce relevant objective functions from recent theoretical findings for chiral electromagnetic fields and employ rigorous simulations of Maxwell's equations.

1. Introduction

Several degrees of freedom of light including frequency, amplitude and polarization are controlled and tailored for applications ranging from optical communication to molecular spectroscopy. Recently, more complex light fields such as occuring in spin-orbit interactions [1] have been investigated. Due to its close relation to these phenomena, the helicity of electromagnetic fields is becoming an important quantity in the analysis of nano-optical devices [2].

Furthermore, nano-antennas have been proven a flexible tool for the enhancement of the emission of electromagnetic sources [3]. In order to increase the contrast of signals from handed sources which possess opposing helicities, it has been shown in the framework of dual symmetry [4] that asymmetric scatterers or metamaterials are required in the nano-optical regime where most materials are non-magnetic.

Often, numerical simulations are carried out for obtaining suitable structures. However, due to the required asymmetry, the underlying parameter spaces are mostly high-dimensional and naive optimization strategy fail to provide fast and reliable results. In addition, fabrication requirements of the devices and/or limitations in the illumination techniques need to be taken into account. That is why, sophisticated optimization algorithms are required [5].

Here, we summarize recent descriptions of the helicity and, in its time-harmonic form, the chirality of light: In section 2, we describe the Helicity Optical Theorem and a Chirality Conservation Law. Precedingly, the stochastical Bayesian optimization [6] is briefly described. Finally, in section 4, we will apply both, the theory on helicity and the described optimization strategy, to the control of two-dimensional line sources embedded in magneto-dielectric dimers.

Content from this work may be used under the terms of the Creative Commons Attribution 3.0 licence. Any further distribution of this work must maintain attribution to the author(s) and the title of the work, journal citation and DOI.
Published under licence by IOP Publishing Ltd

2. Description of optically chiral fields

In standard electrodynamics, dipolar fields are described by the electric $\boldsymbol{p}$ and magnetic $\boldsymbol{m}$ dipole moments. The electric and magnetic fields are up to a prefactor of the wave impedance $Z = \sqrt{\mu/\varepsilon}$ dual to one another [7], where ε is the permittivity and μ is the permeability of the medium in which the source is embedded.

Recently, interest in chiral optical fields has been intensified due to the possibility to enhance the near-field response of chiral molecules. By boosting the optical chirality density locally, extinction measurements gain sensitivity [8]. Additionally, the chirality or helicity of light serves as a tool for fundamental understanding of light-matter interactions [4].

Most optical effects can be described by the electric dipole response. In order to observe chiral effects, the quadrupole contribution of the magnetic dipole response needs to be taken into account. Especially, chiral dipoles of well-defined helicity are of interest in this field. Instead of the dipole basis of electric and magnetic dipoles, the chiral dipole basis of positive and negative handedness with $\boldsymbol{p} = \pm i\boldsymbol{m}$ has been proven suitable for the description of chiral phenomena [9].

2.1. *Helicity Optical Theorem (HOT)*

The response of general dipolar sources $(\boldsymbol{p}, \boldsymbol{m})$ to an incident electric and magnetic field $(\boldsymbol{\mathcal{E}}_{\text{inc}}, \boldsymbol{\mathcal{H}}_{\text{inc}})$ is described by the optical theorem:

$$\frac{\omega}{2}\,\text{Im}\left(\boldsymbol{p}\cdot\boldsymbol{\mathcal{E}}^*_{\text{inc}} + \boldsymbol{m}\cdot\boldsymbol{\mathcal{B}}^*_{\text{inc}}\right) = \mathcal{W}^{\text{abs}} + \frac{ck^4}{3n}\left(\frac{1}{\varepsilon}|\boldsymbol{p}|^2 + \mu|\boldsymbol{m}|^2\right). \tag{1}$$

The left-hand side describes the extinction of energy, $\mathcal{W}^{\text{abs}}$ is the energy absorption and the remaining part is the scattered energy.

In addition to the analysis of energy quantities, a Helicity Optical Theorem has been formulated [2]:

$$\frac{2\pi c}{\mu}\,\text{Re}\left(-\frac{1}{\varepsilon}\boldsymbol{p}\cdot\boldsymbol{\mathcal{B}}^*_{\text{inc}} + \mu\boldsymbol{m}\cdot\boldsymbol{\mathcal{E}}^*_{\text{inc}}\right) = \mathcal{W}^{\text{abs}}_{\mathcal{H}} + \frac{8\pi ck^3}{3\varepsilon}\,\text{Im}\left(\boldsymbol{p}\cdot\boldsymbol{m}^*\right). \tag{2}$$

Here, $\mathcal{W}^{\text{abs}}_{\mathcal{H}}$ is the absorption of helicity (or the conversion of chirality as explained in the next section). The left-hand side is the extinction of chirality and it is obvious that the scattered chirality $\text{Im}\left(\boldsymbol{p}\cdot\boldsymbol{m}^*\right)$ is only non-zero if the electric and magnetic dipole moments are phase-shifted. The chiral dipoles of well-defined helicity mentioned above scatter the magnitude of $\pm|\boldsymbol{p}|^2$ chirality.

This formalism is well suited for the study of extinction experiments with secondary dipolar sources. In the following, we show an equivalent description in the time-harmonic context: the general chirality conservation law. Proceedingly, we apply this law to the analysis of a primary dipolar source.

2.2. *Chirality Conservation Law (CCL)*

The interest in the quantity of optical chirality is manifested in its ability to be described by a general conservation law. This conservation can be formulated for arbitrary, including bi-anisotropic, media. The most general form for monochromatic fields read as [10]

$$2i\omega(\mathfrak{X}_{\text{e}} - \mathfrak{X}_{\text{m}}) + \boldsymbol{\nabla}\cdot\boldsymbol{\mathfrak{S}} = \frac{1}{4}\left\{\boldsymbol{\mathcal{J}}^*\cdot(\boldsymbol{\nabla}\times\boldsymbol{\mathcal{E}}) + (\boldsymbol{\nabla}\times\boldsymbol{\mathcal{J}}^*)\cdot\boldsymbol{\mathcal{E}}\right\}. \tag{3}$$

Similarly to the electromagnetic energy, the chirality density is split into an electric $\mathfrak{X}_{\text{e}} = 1/8\left\{\boldsymbol{\mathcal{D}}^*\cdot(\boldsymbol{\nabla}\times\boldsymbol{\mathcal{E}}) + \boldsymbol{\mathcal{E}}\cdot(\boldsymbol{\nabla}\times\boldsymbol{\mathcal{D}})\right\}$ and magnetic part $\mathfrak{X}_{\text{m}} = 1/8\left\{\boldsymbol{\mathcal{H}}^*\cdot(\boldsymbol{\nabla}\times\boldsymbol{\mathcal{B}}) + \boldsymbol{\mathcal{B}}\cdot(\boldsymbol{\nabla}\times\boldsymbol{\mathcal{H}})\right\}$.

The flux density of optical chirality is $\mathfrak{S} = 1/4\{\mathcal{E} \times (\nabla \times \mathcal{H}) - \mathcal{H}^* \times (\nabla \times \mathcal{E})\}$. The source term on the right-hand side of this continuity equation is of special interest for this study.

Using the current densities $\mathcal{J}$ for electric and magnetic dipoles, (3) yields the HOT in the context of extinction measurements. For the analysis of primary sources, analogous to the case of energy, the conservation of chirality describes the emission X_{emi}, conversion X_{conv} and the far-field accessible chirality flux X_{far}. These quantities are employed for the optimization of chiral effects in the preceeding study and are for localized sources derived as:

$$X_{\text{emi}} = X_0 + \frac{1}{2}\omega \int \text{Im}\left(\mathcal{J}^* \cdot \mathcal{B}_{\text{sca}}\right) d^3r \tag{4}$$

$$X_{\text{conv}} = -2\omega \int \text{Im}\left(\mathfrak{X}_{\text{e}} - \mathfrak{X}_{\text{m}}\right) d^3r \tag{5}$$

$$X_{\text{far}} = \int_{S^2(r\to\infty)} \text{Re}\left(\mathfrak{S}\right) d^2r = X_{\text{emi}} - X_{\text{conv}}. \tag{6}$$

The chirality emission X_{emi} consists of the bulk chirality emission X_0, which is the chirality emitted from the dipole in absence of any interacting environment, e.g. $X_0 = 0$ for achiral sources. In addition, a term occurs in X_{emi} which depends on the current density $\mathcal{J}$, i.e. in our case the dipole moments, and the magnetic field $\mathcal{B}_{\text{sca}}$ scattered by the object. The integration of the chirality flux in the far-field, $r \to \infty$, yields the far-field accessible quantity X_{far}. The chirality conversion X_{conv} consists of a bulk and an interface term for piecewise-constant materials [10].

3. Bayesian optimization

Local optimization is a fairly simple task with various well-established methods such as gradient descent or Newton algorithms. However, in complex systems several optima might arise and mostly only the global optimum is of interest. Bayesian optimization with Gaussian processes tackles this problem and is applied in many fields such as robotics, experimental design and environment monitoring.

Recently, it has also been applied to nano-optical devices [5]. Based on this study, we summarize the basic formalism underlying Bayesian optimization and give details on our implementation in the following.

3.1. Basic formalism

The aim of an optimization is to minimize an objective function f depending on the parameters given by $\boldsymbol{x}$:

$$\tilde{\boldsymbol{x}} = \underset{\boldsymbol{x}\in\mathcal{X}}{\text{argmin}} f(\boldsymbol{x}), \tag{7}$$

where $\tilde{\boldsymbol{x}}$ are the parameters of the global minimum of f in the d-dimensional design space $\mathcal{X} \subset \mathbb{R}^d$. The design space might be a d-dimensional hypercube or a general constraint parameter space.

In each search step Bayesian optimization (BO) takes all previously obtained evaluations of f into account to determine promising parameters and is thus able to reach the global minimum with relatively few iterations. BO is based on a statistical model. Due to their flexibility and tractability, Gaussian processes have been proven to be a suitable basis for this model. Here, every evaluation is associated with a normally distributed random variable.

A Gaussian process is defined by a mean function $\mu : \mathcal{X} \to \mathbb{R}$ and a kernel function $k : \mathcal{X} \times \mathcal{X} \to \mathbb{R}$ which is positive definite and describes the covariance of the process. The probablity of the multivariate Gaussian random variable $\boldsymbol{Y} = [f(\boldsymbol{x}_1), ..., f(\boldsymbol{x}_N)]^T$ is given by

$$P(\boldsymbol{Y}) = \frac{1}{(2\pi)^{N/2}|\boldsymbol{\Sigma}|^{1/2}} \exp\left[-\frac{1}{2}(\boldsymbol{Y}-\boldsymbol{\mu})^T\boldsymbol{\Sigma}^{-1}(\boldsymbol{Y}-\boldsymbol{\mu})\right], \tag{8}$$

where $\boldsymbol{\mu} = [\mu(\boldsymbol{x}_1), ..., \mu(\boldsymbol{x}_N)]^T$ are the mean values and $\boldsymbol{\Sigma} = [k(\boldsymbol{x}_i, \boldsymbol{x}_j)]_{i,j}$ is the covariance matrix.

Based on already known values of the objective function, it is possible to make statistical predictions about not yet evaluated points in the parameter space by Gaussian process regression. The determination of the next evaluation point is based on the predicted Gaussian distribution of the function value (mean and standard deviation). This information is used in the determination of the next evaluation point $\boldsymbol{x}_{N+1}$ by an acquisition function $\alpha : \mathcal{X} \times \mathbb{R} \to \mathbb{R}$:

$$\boldsymbol{x}_{N+1} = \underset{\boldsymbol{x}\in\mathcal{X}}{\text{argmax}}\ \alpha(\boldsymbol{x}, y_{\text{min}}), \tag{9}$$

where $y_{\text{min}} = \min(f(\boldsymbol{x}_1), ..., f(\boldsymbol{x}_N))$ is the currently obtained global minimum.

3.2. Implementation details

Our implementation of BO is based on the Matern-5/2 kernel

$$k(\boldsymbol{x}, \boldsymbol{x}') = \sigma^2 \left(1 + \sqrt{5r^2(\boldsymbol{x}, \boldsymbol{x}')} + \frac{5}{3}r^2(\boldsymbol{x}, \boldsymbol{x}')\right) \exp\left(-\sqrt{5r^2(\boldsymbol{x}, \boldsymbol{x}')}\right) \tag{10}$$

$$\text{with } r^2 = \sum_{i=1}^{d} (x_i - x_i')^2/l_d^2,$$

where σ is the standard deviation and $l_1, ..., l_d$ describe the length scales of the parameters. As noted in [5] this kernel leads to slightly faster optimizations than the frequently used squared exponential kernel.

The hyper-parameters of the Gaussian process $\omega = (\sigma, l_1, ..., l_d)$ are unkown. Since they are at the core of the BO optimization strategy, they should be chosen in an optimal fashion. This optimization is a computationally very expensive part of BO. The employed strategy to reduce the computational overhead of the hyper-parameter optimization is presented in [11].

For the acquisition function α, we use the expected improvement

$$\alpha(\boldsymbol{x}, y_{\text{min}}) = \mathbb{E}[\max_{\boldsymbol{x}}(0, y_{\text{min}} - f(\boldsymbol{x}))], \tag{11}$$

where $f(\boldsymbol{x})$ is the statistical prediction (Gaussian distribution) of the unknown objective function at the position $\boldsymbol{x}$. That is, we take the next sample at the point in the parameter space where the expectation value of the one-sided difference $\max_{\boldsymbol{x}}(0, y_{\text{min}} f(\boldsymbol{x}))$ between the currently known minimum y_{min} and the Gaussian distribution of $f(\boldsymbol{x})$ is maximized.

The BO employed in this study is implemented in *python*. The Gaussian processes are based on the module *gptools*. Further details may be found in [5] and [11].

4. Automatic design of optically chiral dimers for line sources

In this study, we present the application of BO to a magneto-dielectric dimer showing optically chiral behaviour. Different aims and accordingly different objective functions are employed for the optimization of dual symmetry, i.e. conservation of helicity, as well as the selective emission enhancement for one handedness of dipolar chiral sources. Finally, a scatterer is designed preserving one handedness and changing the opposite handed source into an achiral field in the far-field.

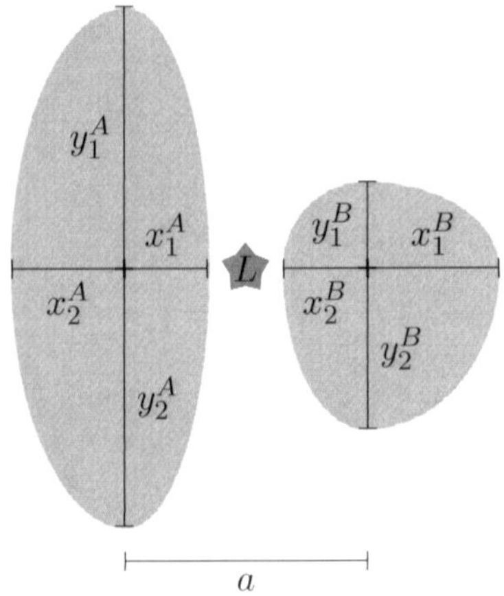

Figure 1. Parametrization of dimer consisting of constitutens A and B each with four different radii x_1, y_1, x_2 and y_2 and separated by distance a. Line Source L is placed in the center of the dimer.

4.1. Setup and parametrization

The scatterer analyzed in the following is a dimer consisting of constitutents A and B (Fig. 1). The centre points of A and B are separated by the distance a. Each constitutent is parametrized by four radii, respectively. This results in a shape given by $(x_i \cos\beta, y_j \sin\beta)^T \subset \mathbb{R}^2$, where x_i and y_j are chosen according to the quadrant and β is the corresponding angle. For example, in the second quadrant: $i = 2, j = 1, \beta \in [\pi/2, \pi]$.

The material of the dimer has a high refractive index of $n = 4$. Accordingly, magnetodielectric effects with high magnetic resonances can be observed, although the material is purely electric with relative permittivity $\varepsilon_r = n^2$ and relative permeability $\mu_r = 1$ [12]. This is crucial for the observation of optically chiral behaviour, since for this application spectrally matched electric and magnetic resonances of similar strength are required.

In the centre of the dimer, we place a line source emitting at wavelength λ. Chiral line sources with $\boldsymbol{p} = \pm i\boldsymbol{m}$ are analyzed. In order to observe chiral effects despite the underlying z-invariance of the setup, we use $\boldsymbol{p} \propto (0,1,1)^T$. The standard Purcell factor of this two-dimensional source differs only in its prefactor from the well-known formula for a three-dimensional dipole [13]:

$$F_P = 1 + \frac{1}{2W_0}\left\{\frac{1}{\omega\mu}\,\mathrm{Im}\left(\boldsymbol{p}^* \cdot \boldsymbol{\mathcal{E}}_{\mathrm{sca}}\right) + k\,\mathrm{Im}\left(\boldsymbol{m}^* \cdot \boldsymbol{\mathcal{H}}_{\mathrm{sca}}\right)\right\}, \tag{12}$$

where W_0 is the energy radiated by the dipole alone, in absence of any interacting environment. The corresponding chirality emission in (4) for a magnetic and electric dipolar sources is [2]

$$X_{\mathrm{emi}} = X_0 + \mathrm{Re}\left(\boldsymbol{p}^* \cdot \boldsymbol{\mathcal{H}}_{\mathrm{sca}}\right) - \frac{1}{Z}\,\mathrm{Re}\left(\boldsymbol{m}^* \cdot \boldsymbol{\mathcal{E}}_{\mathrm{sca}}\right). \tag{13}$$

The electromagnetic response of the devices are computed from rigorous solutions of Maxwell's equations. These are obtained with the Finite Element Method implemented in the commercial Maxwell solver *JCMsuite* [14].

4.2. Optimization of dual elliptical dimers

In order to present the optimization procedure in a low-dimensional space, we fix the emission wavelength $\lambda = 950$nm and the distance $a = 200$nm. Furthermore, we set $x_1 = x_2$ and $y_1 = y_2$

IOP Conf. Series: Journal of Physics: Conf. Series **963** (2018) 012004 doi:10.1088/1742-6596/963/1/012004

for both constitutents A and B of the dimer. In addition, the dimer is mirror symmetric, i.e. $x_A = x_B$ and $y_A = y_B$. This leaves two free parameters, x and y for the optimization. Accordingly, A and B are ellipses and it has been shown that ellipsoids are suitable for matching electric and magnetic resonances [15].

The objective function is chosen as the far-field helicity

$$h = \frac{X_{\text{far}}}{kW_{\text{far}}}, \tag{14}$$

where X_{far} is the chirality and W_{far} is the energy measured in the far-field, respectively. The absolute value of $|h| \leq 1$ is limited to unity. Using a negative chiral source, the helicity $h = -1$ describes so-called *dual* behaviour [9] of the dimer: The chirality observed in the far-field is well-defined and has the same value as the chirality of the emitter.

We use BO to minimize h with respect to the ellipse radii $x \in [20, 80]$nm and $y \in [20, 210]$nm. For 16 parallel evaluations of the objective function, the BO requires 122 evaluations in total to obtain a global minimum of approximately -0.956. This means that only less than 4.4% of the emitted chirality is not preserved. The optimal parameters are $x \approx 68.66$nm and $y \approx 150.64$nm.

As shown in Fig. 2, many different values of h are obtained in the optimization process. The acquired parameters points and their respective objective values are depicted by circles. In addition, Gaussian regression allows for predicting h in the full parameter space and this is shown by the colormap.

In this low dimensional space it is possible for reasonable computational costs to scan the full parameter space which is shown in Fig. 3. In the chosen domain there is only one minimum prominent alongside a larger minimum at the lower corner of the domain. The prediction from BO is very accurate, especially in the surrounding of the global minimum.

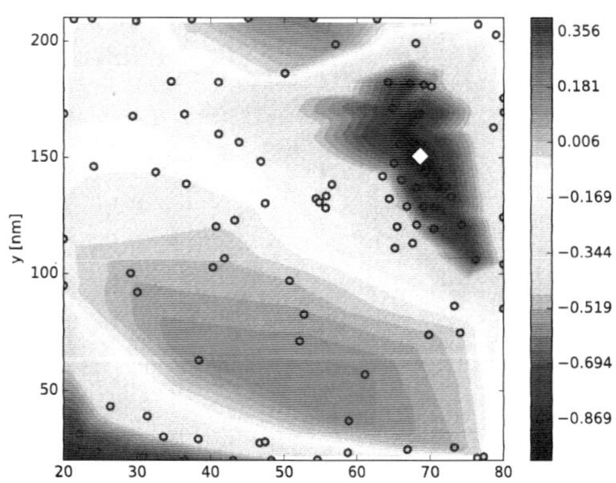

Figure 2. Parameters obtained by BO. Colormap shows Gaussian regression.

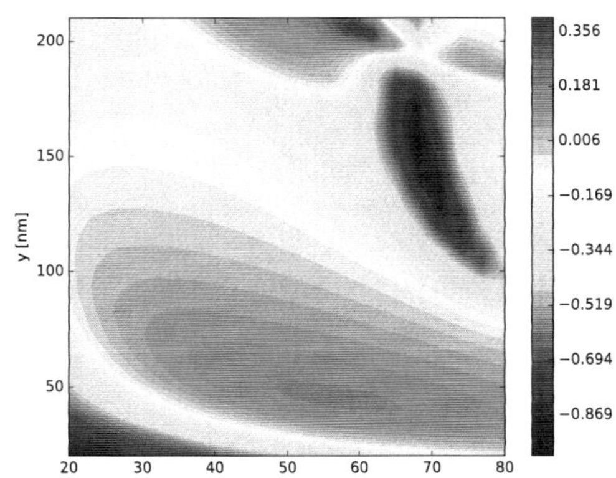

Figure 3. Full parameter space. Colormap shows the objective function.

4.3. Optimization of optically chiral dimers

In the following, we allow optimization of all eight radii and the dimer distance (Fig. 1) resulting in nine free parameters for the BO. Again, we fix the emission wavelength $\lambda = 950$nm. Assuming non-dispersive media, varying the frequency of the emitter requires only scaling of all geometrical parameters of the optimized structure.

The x-parameter space is limited to $[20, 80]$nm and the y parameter space is restricted to $[20, 210]$nm. The distance a is limited to $[40, 350]$nm. Furthermore, the following constraint in the parameter space is employed in order to ensure a minimal distance of the source and the scatterer: $\max(x_1^A, x_2^B) + 20nm - 0.5a < 0$.

For non-mirror-symmetric dimers, it is possible to observe optically chiral effects in the sense that emission of positive and negative chiral sources differ. According to the experimental setup, different quantities might be of interest for the optimization. Here, we optimize for four different objective functions:

- the relative difference of Purcell factors of positive and negative chiral sources

$$f_{g_P}(\boldsymbol{x}) = g_P = 2\frac{F_P^+ - F_P^-}{F_P^+ + F_P^-}, \quad f_{g_P} \in [-2, 2] \tag{15}$$

- the relative difference of Purcell factors ensuring a large F_P^-

$$\widetilde{f}_{g_P}(\boldsymbol{x}) = \left\{\frac{1}{4}g_P + \frac{1}{2}\right\} + \frac{1}{F_P^- + 1}, \quad \widetilde{f}_{g_P} \in (0, 2) \tag{16}$$

- the relative difference of scattered helicity in the far-field

$$f_{g_{\text{far}}}(\boldsymbol{x}) = g_{\text{far}} = |h^+| - |h^-|, \quad f_{g_{\text{far}}} \in [-1, 1] \tag{17}$$

- the relative difference of scattered helicity ensuring a large F_P^-

$$\widetilde{f}_{g_{\text{far}}}(\boldsymbol{x}) = \left\{\frac{1}{2}g_{\text{far}} + 0.5\right\} + \frac{1}{F_P^- + 1}, \quad \widetilde{f}_{g_{\text{far}}} \in (0, 2) \tag{18}$$

Fig. 4 shows the result of the optimization of all nine parameters for the relative difference of Purcell factors f_{g_P}. The sensitivity of the objective value with respect to the three most sensitive parameters a, y_2^B and x_2^B is given. On the x-axis the variation of the parameters in nm and on the y-axis the predicted change of the objective function is shown. These predictions are derived from Gaussian regression and their standard deviation is depicted with dashed lines.

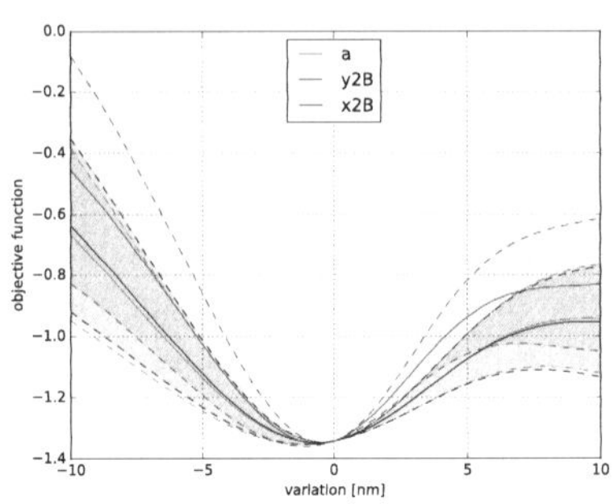

Figure 4. Optimization of f_{g_P}. Most sensitive parameters.

obj. fct.	F_P^-	F_P^+	g_P	opt. geom.
f_{g_P}	8.38	1.65	-1.34	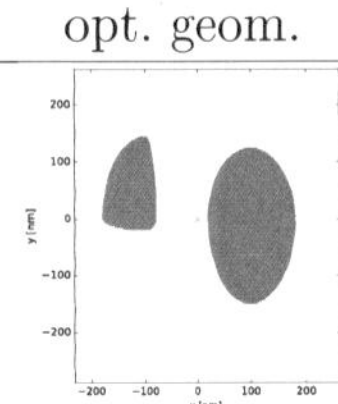
$\widetilde{f}_{g_P}$	9.56	2.09	-1.28	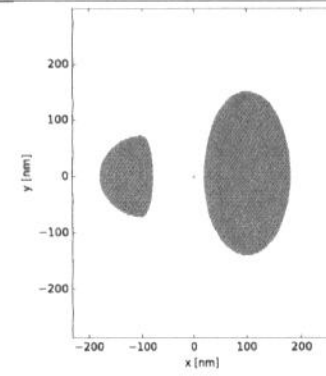

Figure 5. Optimization of f_{g_P} and $\widetilde{f}_{g_P}$. Purcell factors, their relative difference and optimal geometries.

The limits in the parameter space are analogous to the optimization of the dual dimer in the previous section. The detailed geometric parameters are given in Tab. A1. Five out of nine parameters are at the upper or lower limit of the parameter hypercube. A larger parameter space might therefore be beneficial for further optimization, however, larger violation of the dual symmetry is expected.

A minimal g_P-factor of -1.34 is achieved by a highly achiral scatterer. Since the emission of negative chiral sources is larger for a negative g_P-factor, the constitutent B is as close as possible in the limited parameter space to the source which emits mostly to the right. However, a minimized g_P could also be achieved by purely reducing F_P^+. That is why, we change to the objective function to $\widetilde{f}_{g_P}$ which ensures a large F_P^-.

As shown in Fig. 5, the adjusted objective functions results in an increase of more than 10% in the Purcell factor of the negative chiral source F_P^-. On the other hand, g_P is only decreased by less than 5%. These changes in the objective values are rather small. However, it shows that the objective function has to be chosen carefully to obtain the desired behaviour of the device. This will be more important in the next example.

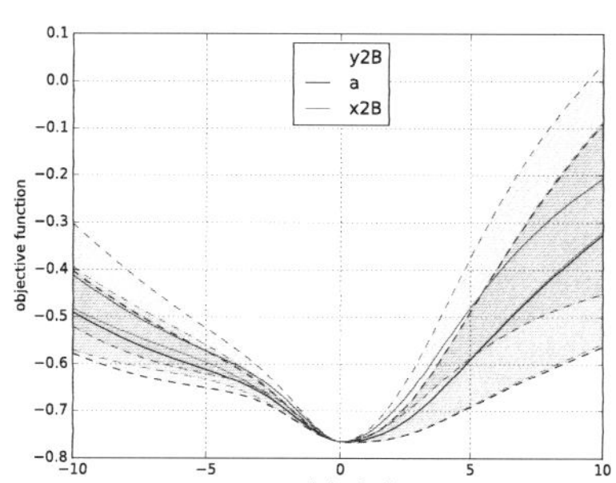

Figure 6. Optimization of $f_{g_{\text{far}}}$. Most sensitive parameters.

obj. fct.	h^+	F_P^-	F_P^+	g_{far}	opt. geom.
$f_{g_{\text{far}}}$	5.3e-6	1.45	0.88	-0.77	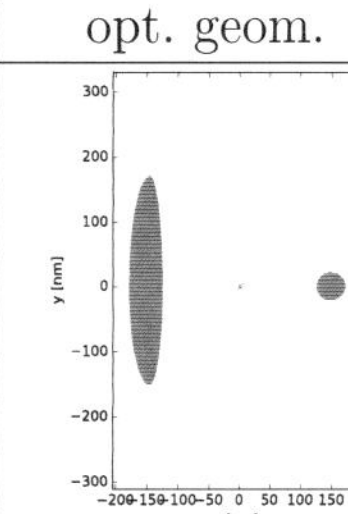
$\widetilde{f}_{g_{\text{far}}}$	-5.5e-3	3.57	1.66	-0.68	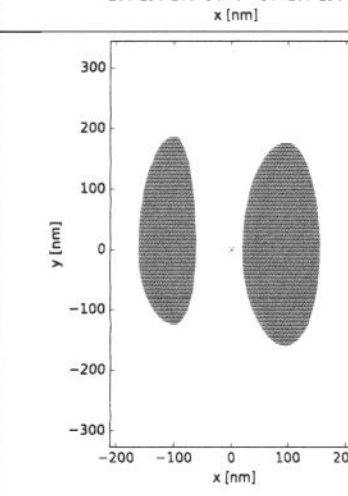

Figure 7. Optimization of $f_{g_{\text{far}}}$ and $\widetilde{f}_{g_{\text{far}}}$. Far-field chiralities, their rel. difference and optimal geometries.

In addition to the standard Purcell factors, it provides new insight to analyze the helicity or chirality of light. Namely, the far-field helicity (14) is of interest in the analysis of e.g. chiral molecules. Note that the investigated structure is lossless and accordingly $W_{\text{far}} = W_{\text{emi}}$, i.e. the Purcell factor F_P is experimentally accessible in the far-field.

Minimizing the objective function $f_{g_{\text{far}}}$ yields a structure which preserves helicity for negative chiral sources and yields achiral emission in the far-field for positive chiral sources. This could be of interest in order to distinguish both handedness with high accuracy. Detailed geometric parameters of the optimal geometry are given in Tab. A2.

As shown in Fig. 6 and Fig. 7, the BO gives an optimal geometry which basically turns the chiral emission of a positive chiral source into achiral fields in the far-field: the helicity is below 1%. On the other hand, chirality of negative chiral sources is dominantly preserved and less

IOP Conf. Series: Journal of Physics: Conf. Series **963** (2018) 012004 doi:10.1088/1742-6596/963/1/012004

than 23% of positive chirality is observed in the far-field. Accordingly, g_{far} of the optimized geometry is well suited to distinguish positive from negative chiral sources by analyzing the chirality observable in the far-field.

However, this comes with the disadvantage of relatively low emission enhancement of $F_P^- = 1.45$. Consequently, we change the objective function from optimizing purely g_{far} to $\widetilde{f}_{g_{\text{far}}}$ which takes an increase in the Purcell factor into account. Fig. 7 shows that the optimal device is very different from the previous result. Instead of minimizing chirality conversion by decreasing the area of the dimer constitutent B to a minimum, the two parts of the dimer are of equal size but different shape.

This approach increases the Purcell factor of the negative chiral source by a factor of nearly 2.5. Nevertheless, g_{far} is decreased to -0.68, still yielding basically achiral far-field behaviour for a positive chiral source but showing higher chirality conversion for negative chiral sources. Since the ratio of F_P^- and F_P^+ is higher for the second optimization, we expect this device having better experimental performance.

In summary, we have shown for standard energy quantities such as the difference of Purcell factors of chiral sources as well as for novel quantities such as the helicity of light that the objective function of the optimization has to be chosen carefully. In the best case, a single objective function is able to take several design goals such as high distinction of positive and negative chiral sources as well as high emission enhancement into account. The choice of different objective function results in geometrically much different devices.

5. Conclusion

We used Bayesian Optimization to optimize the design of two-dimensional dimer structures. Quantities of interest were derived from a general theory of the helicity or chirality of light, namely, the recently introduced Helicity Optical Theorem and Chirality Conservation Law. Even in high-dimensional parameter spaces with nine free parameters, BO is a computationally feasible approach to minimize the number of objective function evaluations and obtain the global minimum in a given parameter space.

We expect that this numerical technique together with the rigorous physical description of optically chiral phenomena, provides further insight into devices for the analysis of chiral phenomena and could give guidelines for experimental realization of measurements of the chirality of light.

Acknowledgments

We acknowledge support by Freie Universität Berlin through the Dahlem Research School and by MINECO-FEDER, grants FIS2012-36113-C03-03, FIS2014-55563-REDC, and FIS2015-69295-C3-1-P.

References

[1] Bliokh K Y, Bekshaev A Y and Nori F 2017 *Physical Review Letters* **119** 073901
[2] Nieto-Vesperinas M 2017 *Phil. Trans. R. Soc. A* **375** 20160314
[3] Abass A, Gutsche P, Maes B, Rockstuhl C and Martins E R 2016 *Optics express* **24** 19638–19650
[4] Fernandez-Corbaton I 2014 *Helicity and duality symmetry in light matter interactions: Theory and applications* Ph.D. thesis Macquarie University, Department of Physics and Astronomy
[5] Schneider P I, Santiago X G, Rockstuhl C and Burger S 2017 *Proc. SPIE 10335* p 103350O
[6] Shahriari B, Swersky K, Wang Z, Adams R P and de Freitas N 2016 *Proc. IEEE* **104** 148–175
[7] Jackson J D 1998 *Classical Electrodynamics* 3rd ed (John Wiley and Sons)
[8] Tang Y and Cohen A E 2010 *Physical review letters* **104** 163901
[9] Zambrana-Puyalto X and Bonod N 2016 *Nanoscale* **8** 10441–10452
[10] Gutsche P, Poulikakos L V, Hammerschmidt M, Burger S and Schmidt F 2016 *Proc. SPIE 9756* p 97560X
[11] Garcia-Santiago X, Schneider P I, Rockstuhl C and Burger S 2017 *in preparation*
[12] Krasnok A, Glybovski S, Petrov M, Makarov S, Savelev R, Belov P, Simovski C and Kivshar Y 2016 *Applied Physics Letters* **108** 211105
[13] Novotny L and Hecht B 2006 *Principles of nano-optics* (Cambridge University Press)
[14] Pomplun J, Burger S, Zschiedrich L and Schmidt F 2007 *physica status solidi (b)* **244** 3419–3434
[15] Lukyanchuk B S, Voshchinnikov N V, Paniagua-Domínguez R and Kuznetsov A I 2015 *ACS Photonics* **2** 993–999

Appendix A. Geometric Parameters

The optimized parameters with f_{g_P} (15) and $\widetilde{f}_{g_P}$ (16) as objective functions are given in Tab. A1. Those with objective functions $f_{g_{\mathrm{far}}}$ (17) and $\widetilde{f}_{g_{\mathrm{far}}}$ (18) are given in Tab. A2.

obj. fct.	x_1^A	y_1^A	x_2^A	y_2^A	x_1^B	y_1^B	x_2^B	y_2^B	a
f_{g_P}	20.00	145.51	80.00	20.00	80.00	124.85	80.00	150.51	200.00
$\widetilde{f}_{g_P}$	20.00	72.62	80.00	71.30	79.99	152.40	80.00	140.15	200.08

Table A1. Optimized geometric parameters (Fig. 1) for optimization of relative difference of Purcell factors (15) and (16).

obj. fct.	x_1^A	y_1^A	x_2^A	y_2^A	x_1^B	y_1^B	x_2^B	y_2^B	a
$f_{g_{\mathrm{far}}}$	20.00	170.00	32.51	150.31	25.62	21.57	20.19	20.00	291.43
$\widetilde{f}_{g_{\mathrm{far}}}$	35.42	186.48	62.75	123.69	56.07	176.39	77.99	159.05	196.44

Table A2. Optimized geometric parameters (Fig. 1) for optimization of relative difference of scattered chirality (17) and (18).

IOP Conf. Series: Journal of Physics: Conf. Series **963** (2018) 012005 doi:10.1088/1742-6596/963/1/012005

Analytical determination of the distributed L and C parameters of metamaterial transmission lines using transmission ABCD matrices representation

E Coskuner and J G Garcia

Grupo de Aplicaciones Electromagnéticas Industriales (GAEMI), Departamento de Ingeniería Electrónica, Universitat Autónoma de Barcelona, Spain

Corresponding author's e-mail address: joan.garcia@uab.es

Abstract. This work point out the possibility to identify the metamaterial behaviour of a transmission line by analysing the matrix elements of the ABCD representation. The method allows to identify the distributed *L*, and *C* of the transmission lines, and therefore the relative permitivity of the metamaterial transmission line, and the characteristic impedance. To ilustrate the Method, a metamaterial periodic structure is analysed using the proposed technique.

1. Introduction

The utilization of metamaterial transmission lines offers the possibility of having additional degrees of freedom for the designers. Since the metamaterial introduction in 1999 by Pendry [1], many efforts have been devoted to developing equivalent circuit models for the design of metamaterial structures [2,3]. Metamaterial equivalent circuit models based on lumped elements are limited to narrowband descriptions, generally around the resonance frequencies, which does not make them very useful from the point of view of the designer. On this regard, the distributed equivalent circuit model results more convenient to describe these structures. This paper establishes an easy way to obtain the distributed L and C parameters of a left-handed transmission lines from measured transmission and reflection parameters (S-parameter, ABCD-parameters or equivalent). The analysis also offers an easy way to identify the metamaterial transmission behavior. Work is in progress to develop an equivalent circuit model based on the distributed L and C identified parameters.

2. Theoretic background

One of the main drawback in many metamaterial transmission lines is the difficult to identify the metamaterial behaviour and the value of their negative characteristic permeability and permittivity. In general, the description of selective high frequency transmission lines is done by using the S-parameters. The design of standard filter response use to be done in terms of these parameters, however, the transmission ABCD matrices offer advantages from the point of view of the identification and characterization of metamaterial devices. Fortunately, the S-parameters and the ABCD representation are univocally linked by analytical expressions [4,5,6]. The ABCD transmission line formalism allows to obtain the transfer matrix of a composite network as the product of the transfer matrices of the cascade subnetworks. By combination the corresponding transmission matrices of the constituent shunt and

Content from this work may be used under the terms of the Creative Commons Attribution 3.0 licence. Any further distribution of this work must maintain attribution to the author(s) and the title of the work, journal citation and DOI.
Published under licence by IOP Publishing Ltd

series elements, the ABCD matrix corresponding to an ideal left-handed transmission matrix of an infinitesimal length *dx* is given by expression (1).

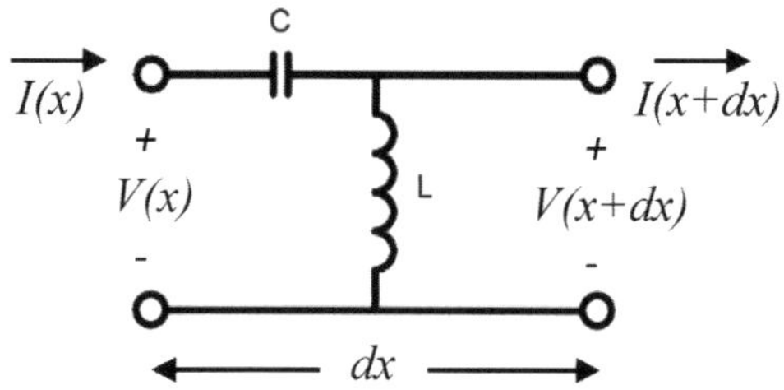

(a)

$$\begin{pmatrix} V(x) \\ I(x) \end{pmatrix} = ABCD_{LH} \begin{pmatrix} V(x+dx) \\ I(x+dx) \end{pmatrix}$$

where (1)

$$ABCD_{LH} = \begin{bmatrix} 1 - \frac{1}{\omega^2 LC} & \frac{1}{j\omega C} \\ \frac{1}{j\omega L} & 1 \end{bmatrix}$$

(b)

Figure 1. (a) Scheme of the Left-handed transmission line and (b) ABCD transmission matrix of a left-handed transmission line infinitesimal section.

where L and C are the distributed inductance and capacitance of the left-handed transmission line. On the other hand, the ABCD matrix of a transmission line can be obtained from the direct measurement or by the transformation of the measured S-parameters. As can be observed in equation (1), the non-diagonal elements of the transmission matrices are negative and pure imaginary, in the case of ideal left-handed transmission lines. From the equation (1) it can also be observed that the ABCD parameters offers a clear method for the identification of the characteristic distributed elements of both right and left-handed the transmission lines. In the case of the left-handed transmission line the non-diagonal elements of the ABCD transmission line allows to evaluate the L and C distributed elements as follow.

$$L = imag \left| \frac{1}{2\pi f C_{LH}} \right| \qquad C = imag \left| \frac{1}{2\pi f B_{LH}} \right| \tag{2}$$

where f is the frequency variable in Hz, L and C are the distributed inductance and capacitance of the line, and C_{LH} and L_{LH} are the corresponding non-diagonal elements of the left-handed ABCD transmission line matrix.

In the case of a discrete line the *dx* in figure 1 should be substituted by the periodicity lattice parameter *a*, equal to the metamaterial unit cell length. The equivalent transfer matrix in this case can be evaluated as the multiplication of the ABCD matrices of the unit cells

$$ABCD_{LH} = ABCD_{UC_1} \cdot ABCD_{UC_2} \cdots = \prod_{i=1}^{N} ABCD_{UC_i} \tag{3}$$

where N is the total number unit cells in the discrete metamaterial structure and $ABCD_{UC}$ is the transmission matrix of the metamaterial unit cell. The resulting transmission matrix, although complicate, keep the main characteristics of the distributed matrix (1). For instance, it can be shown that the elements outside of the resulting matrix $ABCD_{LH}$ in (3) are always purely imaginary, and the diagonal elements are real. This fact, by itself constitute an easy way to check the metamaterial behavior in the left-handed transmission lines formed by periodic structures. Moreover, the analysis of specific cases allows to analytically determine the distributed L, C parameters of the left-handed transmission line as will be shown in the next section. The equation (3) points out the potential complexity of the four elements of the ABCD matrix of a discrete transmission line. In general, each matrix element will be a N-order polynomial fraction. The zeros and the poles of these fraction will introduce divergent points in the transmission line that should tend to disappear in the effective medium limit.

3. Analysis of a left-handed transmission line

Figure 2. (a). shows a periodic structure with three cells between the ports 1 and 2. Each cell is formed by two rectangular patches with a 4mm diameter grounded via centred in the external side of each patch. The unit cell is a variation of the mushroom structure modified to enhance the symmetry in the transmission direction and minimize the loses. The black area corresponds to the top metallic layer of the microstrip structure in a 50mil thick RO3010 substrate. Figure 2. (b) shows the ADS-Momentum simulated response.

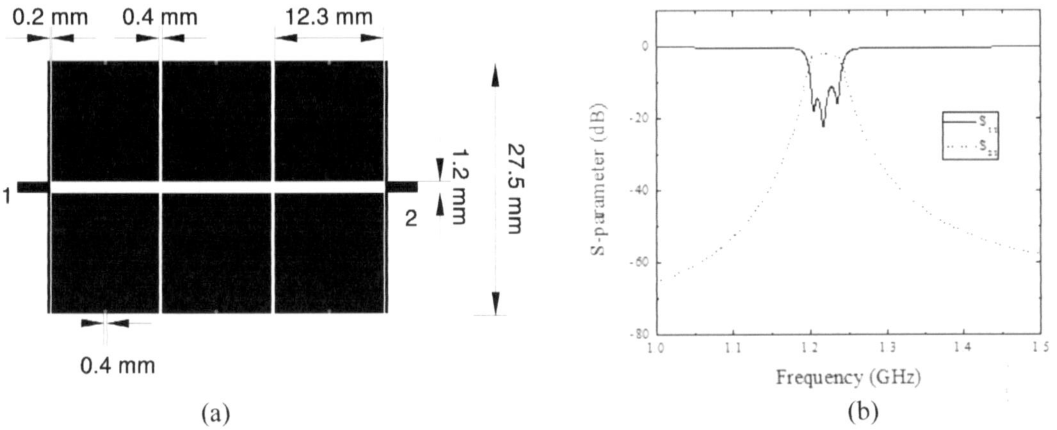

Figure 2. (a) Layout of the microstrip top metallic layer of a left-handed transmission line formed by three-unit cells being *a* the lattice parameter including the more significant dimensions. (b) ADS-Momentum S-parameter simulated response of the LH structure.

The dimensions of the structure showed in figure 2 (a) have been optimized to obtain the bandpass between 1.2 and 1.25 GHz showed in figure 2 (b) Due to the structure of the unit cells of the structure the transmission band is purely left handed, and therefore, in the limit of effective media, it should be described by the equations (1).

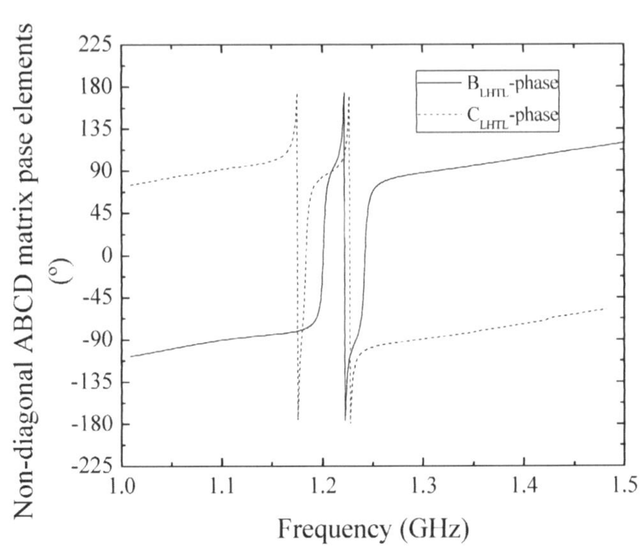

Figure 3. Phase of the non-diagonal elements of the transmission ABCD matrix

The ABCD transmission matrix can be analytically calculated from the S-parameters [3,4,5,6]. Figure 3 shows the phase of the non-diagonal elements of the ABCD transmission matrix of the structure. It can be observed that the phase of these elements is oscillate + and – 90° indicating the complex nature of these none diagonal elements.

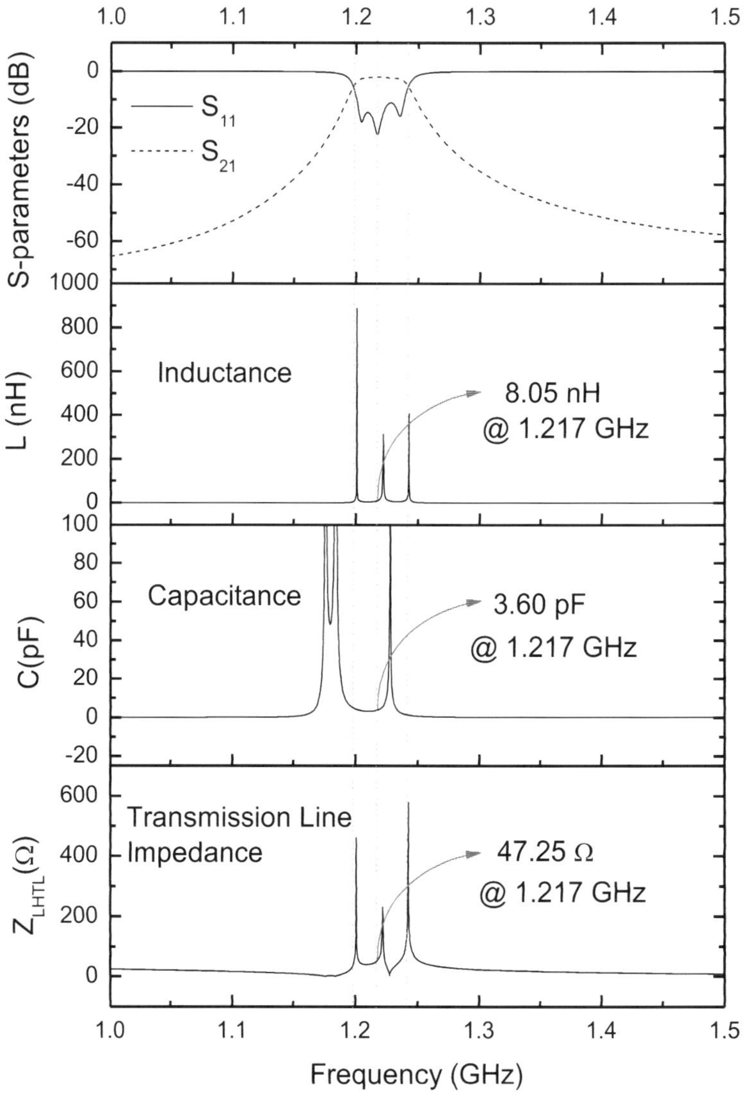

Figure 4. Graphs of the S-parameter simulated response of the structure described in figure 2, calculated total capacitance (C) and inductance (L) of the transmission line and the characteristic impedance of the left-handed transmission line (Z_{LHTL})

Figure 4 shows the result of the proposed analysis applied to the structure described in figure 1. The S-parameters situate a 0.4 GHz bandpass centered at 1.217 GHz. The L and C graphs are obtained applying the expressions (2) to the non-diagonal elements of the ABCD transfer matrix. The results are C=3.60 pF and L=8.05nH for the full length of the transmission line. The average distributed C and L can be

IOP Conf. Series: Journal of Physics: Conf. Series **963** (2018) 012005 doi:10.1088/1742-6596/963/1/012005

evaluated by dividing the value by $3 \cdot a$ since there are three unit cells in the structure showed in figure 1. Finally, the Z_{LHTL} parameter is the characteristic impedance or the left-handed transmission line, obtained as the square root of the L and C ratio.

4. Conclusions

A simple way to identify the left-handed behaviour in a transmission line by the examination of the ABCD matrix elements has been exposed and tested in this work. The transmission ABCD matrix can be analytically obtained from measured S-parameter matrix, and it has been shown how the evaluation of their elements allows to identify the left-handed transmission line distributed *L* and *C* characteristics parameters. To illustrate the method, a mushroom like unidirectional transmission line is analized. Work is in progress to develop distributed equivalent circuit models of the left-handed transmission line based on the distributed parameters evaluated following the exposed method.

'References'

[1] J.B. Pendry, A.J. Holden, D.J. Robbins and W.J. Stewart, "Magnetism from conductors and enhanced nonlinear phenomena", IEEE Transactions Microwave Theory Tech., vol. 47, pp. 2075-2084, November 1999.z

[2] N. Engheta, R. W. Ziolkowski, "Metamaterials Physics and Engineering Esplorations" Wiley & Sons, 2006

[3] F. Capolino, "Metamaterials Handbook. Applications of Metamaterials", CRC Press, 2009.

[4] D. M. Pozar, "Microwave Engineering", John Willey & Sons, Inc. New Cork 1998.

[5] K. C. Gupta, R. Garg, I. Bahl, and P. Bhartis, Microstrip Lines and Slotlines, Second Edition, Artech House, Boston, 1996.

[6] Jia-Sheng Hong & M.L. Lancaster, "Microstrip Filters for RF/Microwave Applications", John Willey & Sons, Inc., New York 2001.

IOP Conf. Series: Journal of Physics: Conf. Series **963** (2018) 012006 doi:10.1088/1742-6596/963/1/012006

Dual circularly polarized broadside beam antenna based on metasurfaces

A Tellechea[1], F Caminita[2], E Martini[2], I Ederra[1,3], J Teniente[1,3], J C Iriarte[1,3], R Gonzalo[1,3] and S Maci[2]

[1] Dept. of Electric and Electronic Eng., Public University of Navarre, 31006 Pamplona, Spain
[2] Dept. of Information Eng. and Mathematics University of Siena, 53100 Siena, Italy
[3] Institute of Smart Cities, Public University of Navarre, 31006 Pamplona, Spain
E-mail: amagoia.tellechea @unavarra.es

Abstract. Design details of a Ku band metasurface (MTS) antenna with dual circularly polarized (CP) broadside radiation is shown in this work. By means of the surface impedance tensor modulation, synchronized propagation of two transversal magnetic (TM) and transverse electric (TE) surface waves (SWs) is ensured in the structure, which contribute to the radiation in broadside direction by the generation of a CP leaky wave. The structure is implemented by elliptical subwavelength metallic elements with a cross-shaped aperture in the center, printed on top of a thin substrate with high permittivity (AD1000 with a thickness of $\lambda_0/17$). For the experimental validation, the MTS prototype has been excited employing an orthomode transducer composed by a metallic stepped septum inside an air-filled waveguide. Two orthogonal TE_{11} modes excited with ±90° phase shift in the feed couple with the TM and TE SWs supported by the MTS and generate RHCP or LHCP broadside beam. Experimental results are compared with the simulation predictions. Finally, conclusions are drawn.

Introduction

In last years, MTS antennas became established as ultrathin, light and easily manufacturable alternative solutions able to meet the restrictive space applications requirements [1-6]. MTS antennas, which are composed by a dense layer of metallic subwavelength elements on top of a grounded dielectric, can be analyzed in terms of the equivalent surface impedance tensor. Its modulation controls the propagation characteristics of a transverse magnetic mode and the generation of a leaky wave with given radiation properties [1]. Isotropic [2] or anisotropic [3-6] structures have been presented in the literature, capable to provide broadside [3], isoflux-shaped [4] or more complex radiation patterns [5] with control on the CP of the field.

Nevertheless, all the structures found in literature provide a single CP radiated field. In [6], the authors presented for the first time the theoretical operating principle of a MTS antenna able to provide RHCP or LHCP broadside beam with a single configuration. This structure deals with the propagation of two SWs of TM and TE nature.

Content from this work may be used under the terms of the Creative Commons Attribution 3.0 licence. Any further distribution of this work must maintain attribution to the author(s) and the title of the work, journal citation and DOI.
Published under licence by IOP Publishing Ltd

Operating principle

As presented in [6], the electric and magnetic tangential fields at each point of a rotationally symmetric MTS supporting the propagation of TM and TE SWs can be related by means of a diagonal surface impedance tensor in cylindrical coordinates, i.e. $\underline{\underline{\mathbf{X}}} = X_{\rho\rho}\hat{\boldsymbol{\rho}}\hat{\boldsymbol{\rho}} + X_{\varphi\varphi}\hat{\boldsymbol{\varphi}}\hat{\boldsymbol{\varphi}}$. The radial ($X_{\rho\rho}$) and azimuthal ($X_{\varphi\varphi}$) tensor components are modulated following a sinusoidal function in radial direction defined by three parameters: the average value (η_{TM}, η_{TE}) normalized with the free space impedance ($\xi = 120\pi$), the modulation index (m_{TM}, m_{TE}) and periodicity (d_{TM}, d_{TE}) related with each mode:

$$\begin{aligned} X_{\rho\rho} &= \zeta\eta_{TM}\left[1+m_{TM}\cos(2\pi\rho/d_{TM})\right] \\ X_{\varphi\varphi} &= \zeta\eta_{TE}\left[1+m_{TE}\cos(2\pi\rho/d_{TE})\right] \end{aligned} \qquad (1)$$

Both modes propagate on the structure balanced in amplitude and synchronized in phase, thus $\eta_{TM}\eta_{TE} = -1$ [6]. Broadside radiation of the generated leaky wave is ensured when the modulation periodicities agree with the surface wave wavelength. Besides, modulation indexes can be appropriately adjusted to ensure balanced contribution to the radiated field. A structure working at 13.5GHz has been implemented with the following parameters: $\eta_{TM} = 1.31$, $m_{TM} = 0.18$, $d_{TM} = 0.6\lambda_0$ related with the TM mode and $\eta_{TE} = -0.76$, $m_{TE} = 0.32$, $d_{TE} = 0.58\lambda_0$ for the TE mode.

Experimental validation

The manufactured MTS antenna is shown in figure 1a. It is composed by a dense layer of subwavelength elliptical metallic patches with a cross-shaped aperture in the centre, printed on top of an Arlon AD1000 grounded substrate with a permittivity of 10.2 and thickness of 1.27mm. At each position of the antenna the patches geometry characteristics and rotation have been selected to provide the surface impedance distribution required for both TM and TE SWs. The structure has been excited with an orthomode transducer composed by a metallic stepped septum inside an air-filled waveguide and a transition from the waveguide to the MTS antenna input port filled with AD1000 dielectric. Two orthogonal TE_{11} modes excited with ±90° phase shift in the feed, couple with the TM and TE SWs supported by the MTS. Depending on the phase shift sign of the excited modes, the generated beam in broadside direction is RHCP or LHCP.

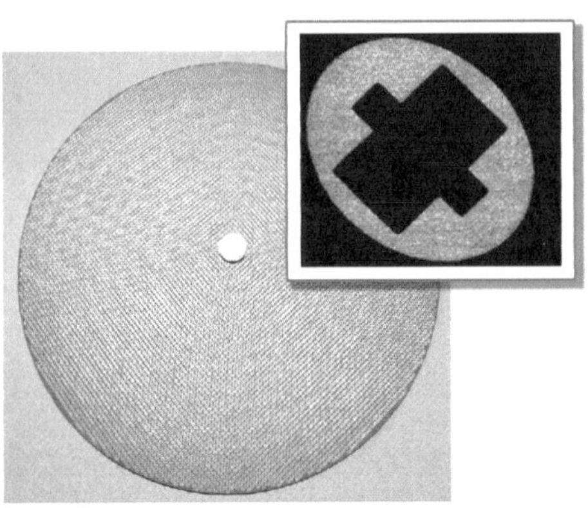

(a)

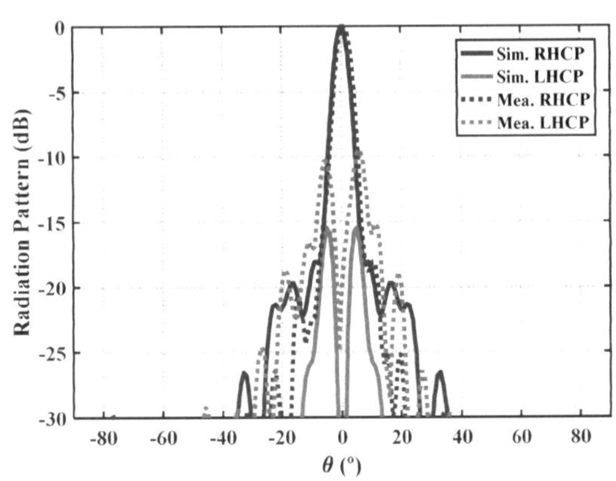

(b)

Figure 1. (a) Manufactured MTS antenna prototype and details of a subwavelength metallic element. (b) Comparison of simulated (continuous lines) and measured (dotted lines) normalized radiation pattern (φ=45°, θ=0°cut) at 13.5GHz: LHCP (red lines) and RHCP (blue lines). The structure has a radius of $8\lambda_0$.

IOP Conf. Series: Journal of Physics: Conf. Series **963** (2018) 012006 doi:10.1088/1742-6596/963/1/012006

Measured radiation pattern at 13.5GHz is compared with simulation predictions in figure 1b. As it can be seen, the structure provides RHCP broadside been at working frequency. Due to the rotational symmetry of the MTS structure, the same behavior is obtained when the feed is excited with 90° phase shift (LHCP).

Conclusions

A MTS antenna with RHCP or LHCP broadside beam radiation at Ku band has been presented. The structure has been implemented by elliptical shaped metallic elements printed on top of a grounded AD1000-based dielectric. Good agreement between the simulation predictions and measurements obtained with the manufactured prototype is observed. More details of the design and manufacturing of the proposed MTS antenna solution and the required feeding system will be presented in the workshop.

References

[1] Fong B H, Colburn J S, Ottusch J J, Visher J L, and Sievenpiper D F 2010 Scalar and tensor holographic artificial impedance surfaces *IEEE Trans. Antennas Propag* **vol 58** pp 3212–3221

[2] Minatti G, Caminita F, Casaletti M, and Maci S 2011 Spiral leaky-wave antennas based on modulated surface impedance *IEEE Trans. Antennas Propag.* **vol 59** pp 4436–4444

[3] Faenzi M, Caminita F, Martini E, De Vita P, Minatti G, Sabbadini M and Maci S 2015 Realization and Measurement of Broadside Beam Modulated Metasurface Antennas *IEEE Antenas and Wireless Propag. Letters* p 1

[4] Minatti G, Maci S, De Vita P, Freni A and Sabbadini M 2012 A circularly-polarized isoflux antenna based on anisotropic metasurface *IEEE Trans. Antennas Propag.* **vol 60** pp 4998–5009

[5] Gonzalez-Ovejero D, Minatti G, Chattopadhyay G and Maci S 2017 Multibeam by metasurface antennas *IEEE Transactions on Antennas and Propagation* **vol 1** pp 1-1

[6] Tellechea A, Caminita F, Martini E, Ederra I, Iriarte J C, Gonzalo R and Maci S 2016 Dual circularly polarized broadside beam metasurface antenna *IEEE Transactions on Antennas and Propagation* **vol 64** pp 2944-2953

Acknowledgments

The authors would like to thank financial support by the Spanish Ministry of Science and Innovation, Direccion General de Investigacion y Gestion del Plan Nacional de I+D+I, Subdireccion General de Proyectos de Investigacion, Project No. TEC2013-47753-C3-1-R and Project No. TEC2016-76997-C3-1-R.

Bottom-up production of meta-atoms for optical magnetism in visible and NIR light

Philippe Barois, Virginie Ponsinet, Alexandre Baron, Philippe Richetti

Centre de Recherche Paul Pascal, CNRS - Université de Bordeaux, 115 Avenue Schweitzer, 33600 Pessac, France.

barois@crpp-bordeaux.cnrs.fr

Abstract. Many unusual optical properties of metamaterials arise from the magnetic response of engineered structures of sub-wavelength size (meta-atoms) exposed to light. The top-down approach whereby engineered nanostructure of well-defined morphology are engraved on a surface proved to be successful for the generation of strong optical magnetism. It faces however the limitations of high cost and small active area in visible light where nanometre resolution is needed. The bottom-up approach whereby the fabrication metamaterials of large volume or large area results from the combination of nanochemitry and self-assembly techniques may constitute a cost-effective alternative. This approach nevertheless requires the large-scale production of functional building-blocks (meta-atoms) bearing a strong magnetic optical response. We propose in this paper a few tracks that lead to the large scale synthesis of magnetic metamaterials operating in visible or near IR light.

1. Introduction: the race towards optical magnetism

Considerable research efforts have been devoted in recent years to control the magnetic permeability of optical materials from GHz to visible light frequencies [1]. It is a difficult challenge, not only because of the well-known absence of optical magnetism in natural dielectrics [2], but also because artificial optical magnetism, generated by induced currents in engineered composite materials, is a complex phenomenon that is not easily related to the familiar permeability parameter [3-5]. The first spectacular success was obtained in 2000 with the realization of the first negative index metamaterial (NIM) based on split-ring magnetic resonators operating in the GHz range. Later on, the observation of strong artificial magnetism at frequencies of visible or near-infrared light was reported in several types of planar plasmonic nanostructures made by top-down technologies and which are essentially two-dimensional and highly anisotropic [6-10]. In addition to the technological complexity linked to the nanometer world, plasmonic resonances suffer from strong optical losses which are likely to hinder future applications. The bottom-up approach is viewed as an alternative pathway that may offer an easier large scale production, including 3-dimensional materials. Furthermore, nanochemistry is able to produce high-refractive index Mie resonators that may usefully replace lossy plasmonic resonators. The aim of this paper is to propose a few tracks for the design and the fabrication of a new generation of optical resonators exhibiting a magnetic response to light.

Content from this work may be used under the terms of the Creative Commons Attribution 3.0 licence. Any further distribution of this work must maintain attribution to the author(s) and the title of the work, journal citation and DOI.
Published under licence by IOP Publishing Ltd

IOP Conf. Series: Journal of Physics: Conf. Series **963** (2018) 012007 doi:10.1088/1742-6596/963/1/012007

2. Plasmonic resonators

One of the first theoretical models of optical magnetic resonators was proposed by Simovski and Tretyakov and is shown in figure 1a [11]. It consists of a corona of metallic nanoparticles surrounding a core dielectric nanoparticle. Light illumination excites a loop of plasmonic currents that act as a magnetic dipole. Such magnetic nanoclusters, sometimes referred to as "plasmonic raspberries", were subsequently synthesized by several authors and their magnetic response was unambiguously demonstrated by optical experiments [12-15].

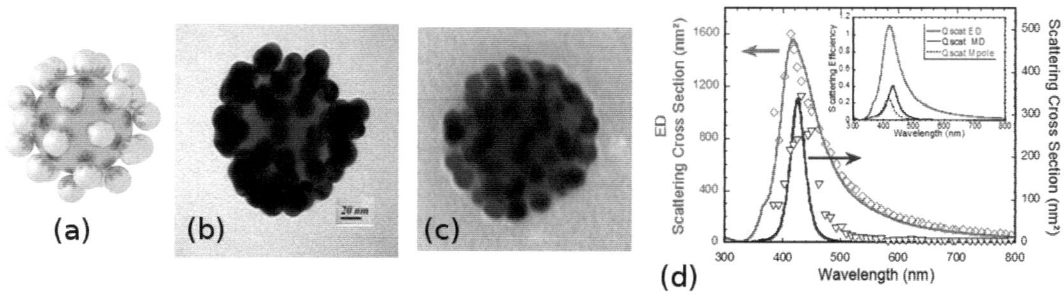

Figure 1: *(a) Sketch of the Simovski-Tretyakov model of magnetic nanocluster. Plasmonic satellites are distributed around a dielectric core nanoparticle. (b-c) Examples of experimental realization with gold (b) and silver (c) satellites. (d) Scattering cross-sections of the magnetic nanocluster shown in (c). Red and blue colors show the contribution of the electric dipole (ED) and of the magnetic dipole (MD) plus electric quadrupole (EQ) respectively. Symbols are the experimental signal, solid lines are numerical simulations. Inset shows the separate contributions of the three scattering processes [15].*

The Simovski-Tretyakov magnetic nanoclusters were subsequently assembled in a three-dimensional metamaterial in a micro-evaporator engineered on a microfluidic chip [16,17]. The effective magnetic permeability μ of the metamaterial was extracted from variable angle spectroscopic ellipsometry. Two important results were obtained for the first time namely (i) the real part of μ deviates significantly from the natural value $\mu = 1$ (in μ_0 units) as $0.8<\mu<1.4$, hence breaking the well-known Landau-Lifshitz rule [2]. The strength of the magnetic response is more obvious if one notes that the effective magnetic susceptibility $\chi = \mu - 1$ is 3 to 5 orders of magnitude larger than the diamagnetic susceptibility of natural materials and (ii) a constant μ parameter describes the reflection from the metamaterial at all angles of incidence, hence showing that the measured permeability is not affected by spatial dispersion, a non-local phenomenon which usually precludes a valid description of optical magnetism by the sole permeability tensor [4,5]. These two results of fundamental importance follow from the hierarchical self-assembly process whereby the meta-atoms are engineered in a first step to provide the wanted magnetic response and assembled in a second step to produce a bulk sample. Far from being a disadvantage, the highly-symmetric liquid-like structural disorder of the metamaterial turns out to be highly favorable for the limitation of the spatial dispersion effects [15]. The bottom-up approach hence appears as a powerful method for the fabrication of magnetic metamaterials exhibiting a non-natural, but still effective magnetic permeability parameter.

The magnetic response of the plasmonic raspberries is however still too low to warrant useful applications. Indeed, the interesting domains μ near zero and negative μ are not reached yet. Although numerical simulations show that increasing the number and the size of the metallic satellites would increase the magnetic response of the plasmonic raspberries, the presence of Ohmic losses in metals undeniably constitutes a severe limitation for future applications in optical devices. All-dielectric meta-atoms exhibiting Mie resonances appear as a promising alternative [18].

3. All-dielectric Mie resonators

The absorption and scattering of light by a homogeneous spheres was worked out exactly by Gustav Mie in 1908 [19]. A remarkable feature is the presence of cavity-type resonances when the diameter of the sphere D matches the wavelength λ_0/n_{sphere} of light inside the sphere. At lowest dipolar order, two types of resonance are actually found namely an electric resonance for $D \approx 2\lambda_0/n_{sphere}$ and a magnetic one for $D \approx \lambda_0/n_{sphere}$. These two resonances are characterized by the radiation of a strong electric and magnetic dipole respectively, the latter being of crucial importance for the generation of optical magnetism. For a dielectric sphere and for wavelengths far from any absorption band of the dielectric material, optical losses are weak and the figure of merit of the resonances is very high. Mie resonators are thus expected to be far more efficient than plasmonic ones [18]. If effective optical parameters are wanted however, the homogeneity condition requires that the size D of the resonators and the distance between them should be much shorter than the wavelength, which is inconsistent with the resonance conditions, unless the refractive index of the material satisfies $n_{sphere} >> 1$. For visible light, D should not exceed a few tens of nanometers whereas a few hundreds may be acceptable for IR waves.

For material scientists, the challenge is therefore to synthesize nanoparticles with the following specifications:

(i) the dielectric material should have a refractive index with a large real part (ideally $n_{sphere} \sim 10$) and low imaginary coefficient k.

(ii) the size of the nanoparticles should be accurately controlled since the resonance condition depends critically on size.

(iii) the synthetic route should be able to produce large amounts of particles in environment-friendly processes.

These are very tough challenges which are not fully solved yet. Several types of dielectric materials are under investigation, among which crystalline silicon (c-Si), amorphous silicon (a-Si) and titanium dioxide (TiO_2) in its crystalline anatase form seem quite promising. Figure 2 below shows that the magnetic dipolar scattering efficiencies calculated from c-Si, a-Si and TiO2 spheres of diameter D = 150 nm are much higher than the value obtained for plasmonic raspberries of the same size [21].

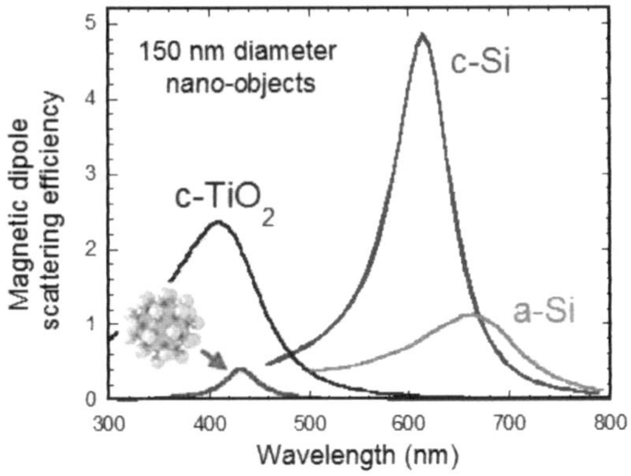

Figure 2: *Scattering efficiencies of the magnetic dipole computed for different nanospheres of diameter 150 nm, compared to the value measured on plasmonic raspberry of the same size (red line).*

For visible light applications, crystalline silicon (c-Si) appears as the most favorable material, although not ideal. Its refractive index is around 4 and the imaginary part is only a few 10^{-2} for wavelengths longer than 500 nm [21]. Figure 2 shows the evolution of the electric and magnetic scattering efficiencies of c-Si nanoparticles with diameter, as calculated from the Mie theory.

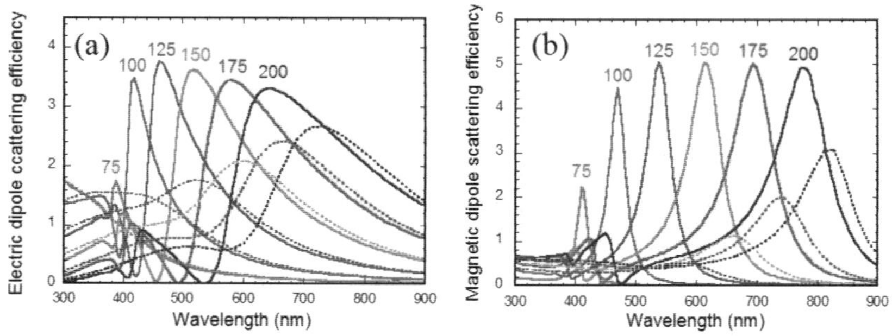

***Figure 3**: Electric (a) and magnetic (b) scattering efficiencies of crystalline silicon nanoparticles of various diameters (calculated from Mie theory in ethanol solvent of index 1.36).*

The electric and magnetic resonances have indeed been observed and characterized in full agreement with the Mie theory on c-Si nanoparticles prepared by laser ablation [22-24]. These studies confirm that the crystallinity of silicon is critical. Although laser fabrication is able to prepare 2D arrays of c-Si particles of well controlled size [23], the implementation of a robust synthetic route able to produce macroscopic amounts (grams to kilograms) of c-Si nanoparticles at low cost would be welcome.

Despite ongoing efforts of nanochemists, c-Si nanoparticles of well-defined size in the range 50-200 nm are not yet available. However, large amounts of silicon nanoparticles are commercially available with a lower degree of crystalline purity and a broad distribution of sizes. Figure 3 shows the result of a light scattering study carried out on silicon nanoparticles purchased from American Elements [26]. The optical signal plotted in Fig. 3c is the ratio of the transverse to axial signals collected at 90° scattering angle along polarizations respectively perpendicular and parallel to the scattering. In the spectral region of the dipolar Mie resonances, it identifies with the ratio of the dipolar magnetic to dipolar electric scattering cross sections [14,27]. The measurement is fully consistent with a population of silicon nanoparticles with a log-normal distribution of size of average diameter 100 nm and a 40% crystallinity. Diffraction of electrons in TEM studies confirm the presence of crystalline and amorphous nanospheres in the commercial silicon powder. Despite a broad size distribution (+/- 20 nm in radius) and a low degree of crystallinity, the ratio of the magnetic to electric scattering reaches 0.6, whereas the maximum value for plasmonic raspberries does not exceed 0.28 [15].

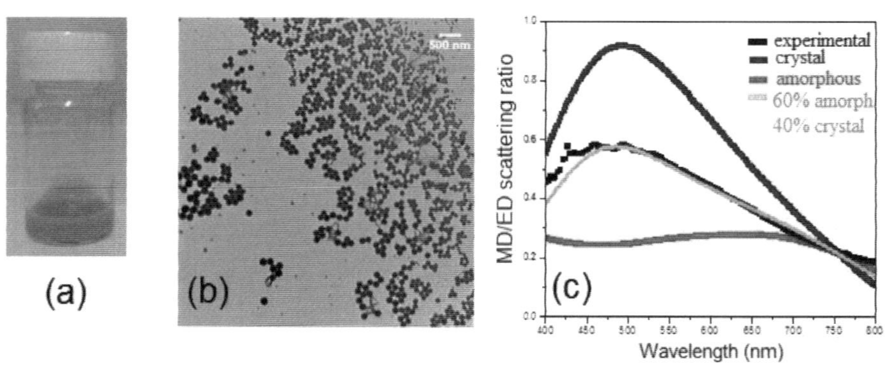

***Figure 4:** (a) suspension of silicon nanoparticles purchased from American Elements in water. (b) TEM micrograph of the Si nanoparticles after 30 minutes exposure to NaOH at pH=12 to eliminate non-spherical oxide growth. A log-normal size distribution of average diameter 100 nm was extracted from TEM images. (c) Plot of the ratio of the magnetic (MD) to electric (ED) dipolar scattering cross sections. Experimental signal in black is well reproduced by a mixture of 40% crystalline and 60% amorphous particles (green line). Blue and red lines represent the signal computed with the same size distribution for 100% crystalline and 100% amorphous particles respectively.*

These encouraging results show that silicon nanoparticles produced in large amounts and low cost by chemical companies can be a good starting point for optical applications. Narrower size distributions can in principle be obtained by powerful methods of physical chemistry like centrifugation in viscosity gradients [28], whereas thermal annealing may improve the crystalline fraction.

4. Conclusion:

The bottom-up route based on self-assembly of simple nano-sized building blocks has proved its efficiency in the production of metamaterials of large area and/or large volume. Structural disorder is not always detrimental for optical functionality, although it makes numerical simulations undeniably more difficult. The main challenge is the design and the large-scale availability of building units of well controlled morphology. Metallic oxide nanospheres such as TiO_2 are already commercially available with a good control of size and crystalline purity. Silicon nanospheres of well controlled size may become available in large amounts in a near future.

Acknowledgements:
This work was supported by the LabEx AMADEus (ANR-10-LABX-42) in the framework of IdEx Bordeaux (ANR-10-IDEX-03-02), France.

References

[1] Soukoulis, C. M., and Wegener, M. Past achievements and future challenges in the development of three-dimensional photonic metamaterials. *Nature Photon.*, **5,** 523-530 (2011).
[2] Landau, L. D. & Lifshitz, E. M. Electrodynamics of Continuous Media, Pergamon Press, 1960).
[3] Agranovich, V. M., and Ginzburg V. L. *Crystal Optics with Spatial Dispersion, and Excitons*. Cardona, M., Fulde, P. and Queisser, H.-J. editors. Volume 42. (Springer-Verlag. Berlin, Heidelberg, 1984).
[4] Agranovich, V. M., and Gartstein, Y. N. Spatial dispersion and negative refraction of light. *Phys. Usp.*, **49,** 1029-1044 (2006).
[5] Menzel, C., *et al.*, Validity of effective material parameters for optical fishnet metamaterials. *Phys. Rev. B*, **81,** 035320 (2010).
[6] Grigorenko, A. N., *et al.* Nanofabricated media with negative permeability at visible frequencies. *Nature*, **438,** 335-338 (2005).
[7] Linden, S. *et al.*, Magnetic Response of Metamaterials at 100 Terahertz. *Science*, **306,** 1351-1353 (2004).
[8] Zhou, J., Zhang, L., Tuttle, G., Koscny, Th., and Soukoulis, C. M. Negative Index materials using simple short wire pairs, *Phys. Rev. B*, **73,** 041101(R) (2006).
[9] Dolling, G., Wegener, M., Soukoulis, C. M. & Linden S. Negative-index metamaterial at 780nm wavelength. *Optics Lett.*, **32,** 53-55 (2007).
[10] Xu, T., Agrawal, A., Abashin, M., Chau, K. J., and Lezec, H. J. All-angle negative refraction and active flat lensing of ultraviolet light. *Nature*, **497,** 470-474 (2013).
[11] Simovski C.R. and Tretyakov S.A., Phys. Rev. B **79**, 045111 (2009).
[12] Mühlig S., Cunningham A., Scheeler S., Pacholski C., Bürgi T., Rockstuhl C., and Lederer F., ACS Nano **5**, 6586 (2011).
[13] Sheikholeslami S.N., Alaeian H., Koh A.L., and Dionne J.A., Nano Lett. **13**, 4137 (2013).
[14] Ponsinet *et al.*, Phys. Rev. B **92**, 220414(R) (2015).
[15] Gomez-Graña *et al.*, Mater. Horiz. **3**, 596 (2016).
[16] Leng J., Lonetti B., Tabeling P., Joanicot M. and Ajdari A., Phys. Rev. Lett., **96**, 084503 (2006).

[17] Angly J. *et al.*, ACS Nano, **7**, 6465 (2013).
[18] Jahani S., Jacob Z., *Nat. Nanotechnol.* **11**, 23 (2016).
[19] Mie G., Ann. Phys. **25**, 377 (1908).
[20] Bohren C.F., Huffman D.R., *Absorption and Scattering of Light by Small Particles*, Wiley-Interscience, New York, **1983**.
[21] Values of the refractive indices are found in https://refractiveindex.info/
[22] Evlyukhin A.B. *et al.*, Nano Lett., **12**, 3749 (2012).
[23] Zywietz U., *et al.*, Nat. Commun., **5**, 3402 (2014).
[24] Dmitriev P.A. *et al.*, Nanoscale, **8**, 5043 (2016).
[25] L. Shi, J. Harris, R. Fenollosa, I, Rodriguez, X. Lu, B. A. Korgel, F. Meseguer, *Nat. Commun.* **2013**, 4, 1904
[26] https://www.americanelements.com/
[27] Sharma N.L., Phys. Rev. Lett. **98**, 217402 (2007).
[28] Qiu P. & Mao C., Adv. Mater., **23**, 4880 (2011).

Application of Gielis transformation to the design of metamaterial structures.

C. M. de Jong van Coevorden†, J. Gielis†,‡ and D. Caratelli†*

†The Antenna Company, High Tech Campus, 41, 5656AE - Eindhoven, TheNetherlands
‡University of Antwerpen, Groenenborgerlaan 171, B2020 - Antwerp, Belgium
*Tomsk Polytechnic University, Sovetskaya Street 84/3, 634050, Tomsk, Russia

carlos.vancoevorden@antennacompany.com

Abstract. In this communication, the use of Gielis transformation to design more compact metamaterial unit cells is explored. For this purpose, transformed complementary split ring resonators and spiral resonators are coupled to micro-strip lines and theirbehaviour is investigated. The obtained results confirm that the useof the considered class of supershaped geometries enables the synthesis of very compact scalable microwave components.

1. Introduction

GielisTransformations (GT) [1] aregeometric transformations acting on planar functions f(θ) generalizing the Lamé equation that unifies a broad variety of natural and abstract shapes (Eq. 1). GTs can morph curves, such as circles, spirals or trigonometric functions, into infinite number of shapes, including regular polygons (Fig.1 a-c), starfish-like shapes (Fig. 1 d-f) and more complexspirals (Fig. 1 g-i).

$$k(\theta, a, b, m_1, m_2, n_1, n_2, n_3) = f(\theta)\left[\left|\frac{1}{a}\cos\left(\frac{m_1}{4}\theta\right)\right|^{n_2} + \left|\frac{1}{b}\sin\left(\frac{m_2}{4}\theta\right)\right|^{n_3}\right]^{n_1} \quad (1)$$

Since its introduction in the literature two decadesago, GTs have been extensively used in many fields of Science and Engineering for addressing a broad range of applications such as the solution ofboundary value problems [2, 3], pattern recognition [4], optimization of heat-shields in manned space vehicles [5], as well asin applied electromagnetics, for example in antennas [6, 7, 8, 9] and filterdesign [10].

The progressively more intensive use ofuse of GTs in engineering is mainly due to theireasy implementation in computer simulation tools, flexibility in design, as well asthe existence offamilies of shapes with specialproperties such as the class of closed curves featuring constant area but different lengths. These properties make GTs an ideal tool to study the effect of geometrical characteristics on the physical response of complex systems.

One meaningful examplein which the geometrical propertiesof the system play a main role in the definition of the relevant physical properties is related to metamaterials. Inthese type of structures a

Content from this work may be used under the terms of the Creative Commons Attribution 3.0 licence. Any further distribution of this work must maintain attribution to the author(s) and the title of the work, journal citation and DOI.
Published under licence by IOP Publishing Ltd

IOP Conf. Series: Journal of Physics: Conf. Series **963** (2018) 012008 doi:10.1088/1742-6596/963/1/012008

unit cell, in general resulting from the combination of different materials, e.g. conductors and dielectrics,is arranged periodically along a suitable lattice in such a way that new properties, such as negative-valued effective permittivity and/or permeability, emerge.This communication explores the application of GTs to two classes of metamaterial unit cells widely adopted in microwave filter design, namely complementary split ring resonators in Section 2 and spirals in Section 3.

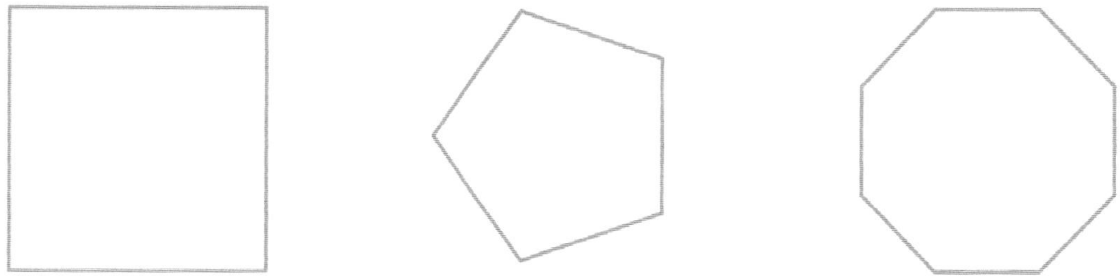

a) $a; b = 1$, $m_1,m_2 = 4$, $n_1 = 5000$, $n_{2,}; n_3=1150$ and $f[\theta] = 1, 0 \leq \theta \leq 2\pi$

b) $a; b = 1$, $m_1;m_2 = 5$, $n_1 = 5000$, $n_2; n_3 = 1150$ and $f[\theta] = 1, 0 \leq \theta \leq 2\pi$

c) $a; b = 1$, $m_1;m_2 = 8$, $n_1 = 5000$, $n_2; n_3 = 1150$ and $f[\theta] = 1, 0 \leq \theta \leq 2\pi$

d) $a; b = 1$, $m_1;m_2 = 4$, $n_1 = 10$, $n_2; n_3 = 30$ and $f[\theta] = 1, 0 \leq \theta \leq 2\pi$

e) $a; b = 1$, $m_1;m_2 = 5$, $n_1 = 10$, $n_2; n_3 = 30$ and $f[\theta] = 1, 0 \leq \theta \leq 2\pi$

f) $a; b = 1$, $m_1;m_2 = 8$, $n_1 = 10$, $n_2; n_3 = 30$ and $f[\theta] = 1, 0 \leq \theta \leq 2\pi$

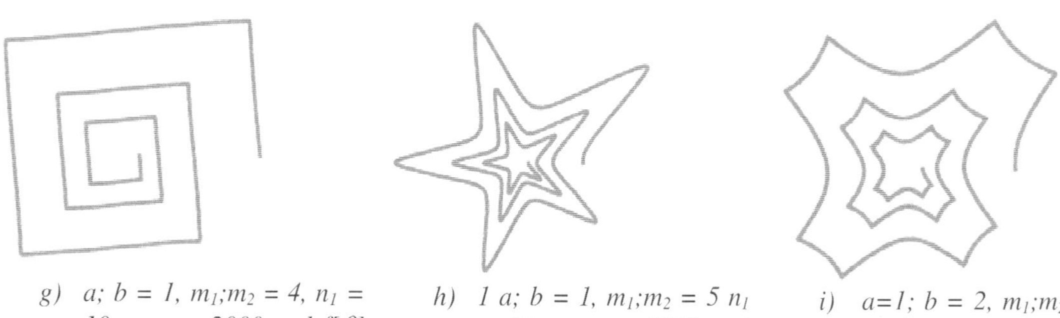

g) $a; b = 1$, $m_1;m_2 = 4$, $n_1 = 10$, $n_2; n_3=2000$ and $f[\theta] = e^{-0.1\,\theta}$, $0 \leq \theta \leq 6\pi$

h) 1 $a; b = 1$, $m_1;m_2 = 5$ $n_1 = 10$, $n_2; n_3 = 2000$ and $f[\theta] = e^{-0.1\,\theta}$, $0 \leq \theta \leq 6\pi$

i) $a=1; b = 2$, $m_1;m_2 = 8$, $n_1 = 600$, $n_2; n_3 = 400$ and $f[\theta] = e^{-0.1\,\theta}$, $0 \leq \theta \leq 6\pi$

Figure: 1 Transformations of a circle into regular polygons (a-c) and starfishes (d-f) and transformations of logarithmic spirals (g-i)

2. Gielis transformed Complementary split ring resonators

The use of split ring resonators (SRRs) and their complementary variant (CSRRs) as basic resonant units in planar microwave filters is being increasingly adopted due to their compact footprint compared to conventional resonators. This enables the realization of semi-lumped filtering structures with high performance and controllable characteristics [11, 12]. Typically, circular or square geometries are chosen for the design of the considered class of resonators, although other recent approaches use fractal or quasi-fractal geometries for the realization of miniaturized CSRRs, but still with a square-like form factor [13, 14].

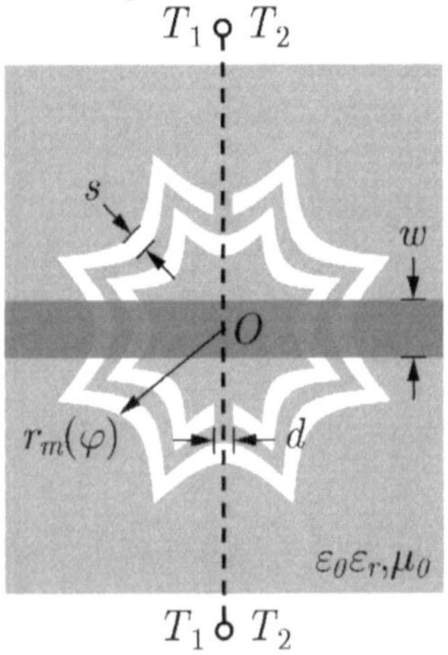

a) Topology where s = d = 0.75 mm on a substrate with ε_r = 3.78 and tanδ = 0.025;

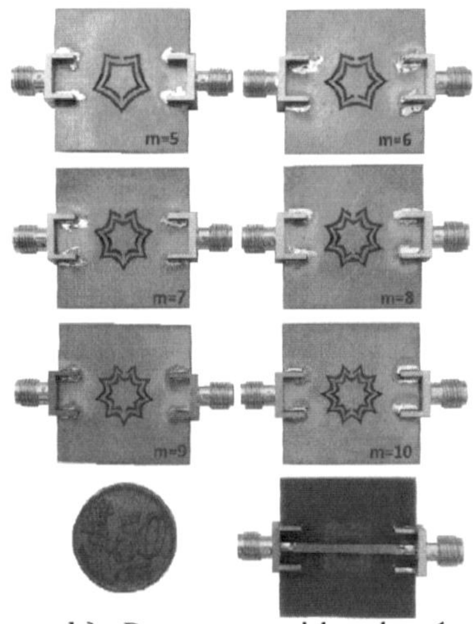

b) Prototypes with a; b = 1, n_1; n_2; $n_3 \to \infty$ and m = 5, 6, 7, 8, 9, 10.

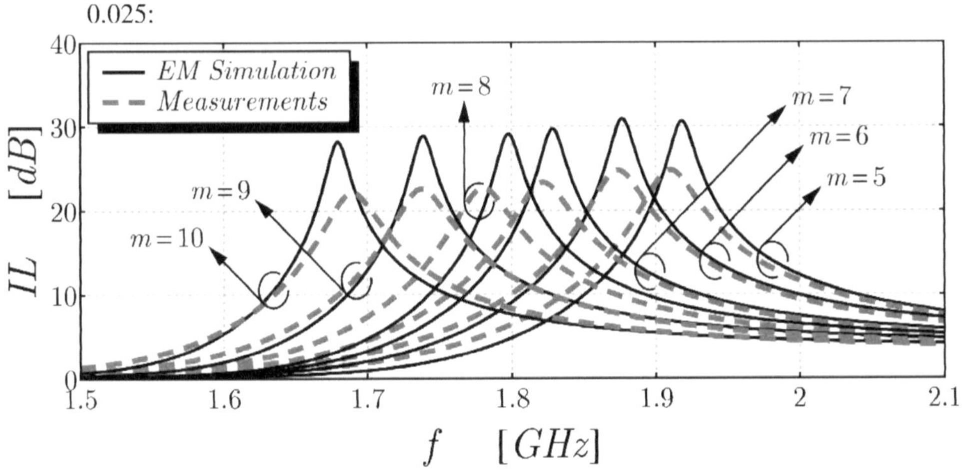

c) Comparison between simulated and measured insertion loss of the Gielis transformed CSSR single cell filter.

Figure: 2 Gielis transformed CSRR-based filters realized in microstrip technology.

Since resonance of CSRRs depends on the average length of the frequency slots and the area of the unit cell [15] one possible way to miniaturize CSRRS is to increase the length of the ring. A well-known property of GT of circles is toincrease length by increasing the symmetry parameter *m* andkeep the rest of parameters unchanged.The enclosed area is then invariant. Making use of this property, the results on the band-stop properties of a microstrip line loaded with GT transformed CSRRs is shown in Figure 2a. Several prototypeswith increasing parameter *m* have been manufactured (Fig. 2b) and their insertion loss measured (Fig.2c). These results show that the electrical size of the unit cell gets larger with *m* as expected, based on the mathematical properties of GT.It is worth noting that, despite the size reduction, the fractional bandwidth is not significantly compromised.

3. Gielis transformed logarithmic spirals.

Spiral resonators (SRs) are increasingly used as unit cells for metamaterials due to their furtherreduction of the electrical size. Following the same development trendsas for SRRs andCSRRs, the first SRs were based on canonical shapes [16], while later other more complexconceptsbased on fractals were also tested [17, 18]. To demonstrate the miniaturization capabilities ofGTs spirals, a comparison of a circular spiral, a fractal spiral as in [18] and a GT spiral hasbeen made with same maximum radius r_{max} (see Fig. 3a). The topology and the prototypes are shown in Figure 3a and 3b respectively, and simulation results are shown inFigure 3c.

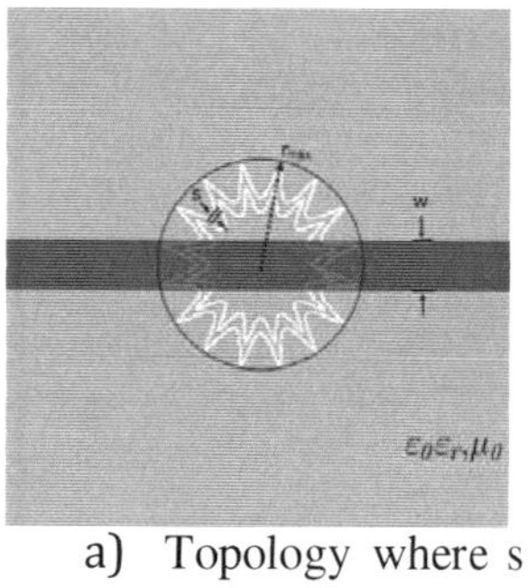

a) Topology where s = d = 0,2 mm on a substrate with ε_r = 4.35 and tanδ = 0.035.

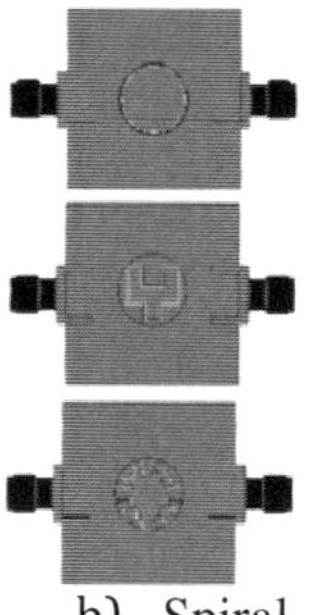

b) Spiral circular (top), Fractal [18] (center). and Gielis transformed spiral (bottom)

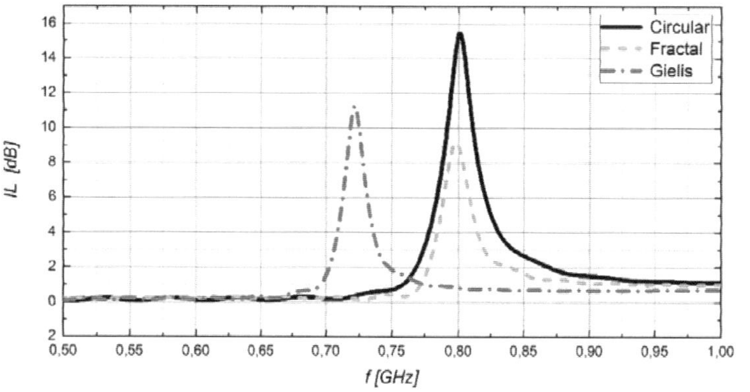

c) Simulated insertion loss of circular, fractal and GT spirals.

Figure: 3. Simulated insertion loss of circular, fractal and GT spiral.

The results displayed in Fig3.c show that transforming canonical spirals with GTs, electrically smaller unit cells can be designed without compromising bandwidth.

4. Discussion

GT have been used for metamaterials in the visible range [19], achieving double negative refractive index with supershaped unit cells, optimized via evolutionary algorithms on all parameters to operate in the blue region of the visible light spectrum. In this contribution, we show how the mathematical invariants of GT can be exploited to even reduce the electrical size of CSRRs and SRs unit cells. These structures show a reduction of around12% of the resonance frequency for the same maximum radius, and close to 30% when the same unit cell area is considered. These results are particularly relevant if the metamaterial has to be physically described as a continuous medium rather than as a discrete periodic structure.

References

[1] J. Gielis. A generic geometric transformation that unifies a wide range of natural and abstract shape, AmericanJournal of Botany.
[2] J. Gielis, D. Caratelli, Y. Fougerolle, P. E. Ricci, I. Tavkelidze, and T. Gerats, Universal natural shapes: From unifying shape description to simple methods for shape analysis and boundary value problems," PLOS ONE, vol. 7, pp. 1-11, 09 2012.
[3] P. Natalini, R. Patrizi, and P. E. Ricci, The Dirichlet problem for the Laplace equation in a starlike domainof a Riemann surface,Numerical Algorithms, vol. 49, pp. 299-313, Dec 2008.
[4] Y. Fougerolle, J. Gielis, and F. Truchetet, A robust evolutionary algorithm for the recovery of rational Gielis curves, vol. 46, pp. 2078 - 2091, 08 2013.
[5] J. E. Johnson, R. P. Starkey, and M. J. Lewis, Aerodynamic stability of reentry heat shield shapes for acrew exploration vehicle," Journal of Spacecraft and Rockets, **2066.**
[6] A. G. Koutinos, G. A. Ioannopoulos, M. T. Chryssomallis, G. A. Kyriacou, and D. Caratelli, Bandwidthenhancement of a supershape patch antenna using multiple feeding technique, in 2017 International Workshop on Antenna Technology: Small Antennas, Innovative Structures, and Applications (iWAT), pp. 211-214, March 2017.
[7] L. Mescia, P. Bia, D. Caratelli, M. A. Chiapperino, O. Stukach, and J. Gielis, Electromagnetic mathematicalmodeling of 3Dsupershaped dielectric lens antennas,Mathematical Problems in Engineering, 2016.
[8] P. Bia, D. Caratelli, L. Mescia, and J. Gielis, Analysis and synthesis of supershaped dielectric lens antennas,, IET Microwaves, Antennas Propagation, vol. 9, no. 14, pp. 1497-1504, 2015.
[9] M. Simeoni, R. Cicchetti, A. Yarovoy, and D. Caratelli, Plastic-based supershaped dielectric resonatorantennas for wide-band applications,IEEE Transactions on Antennas and Propagation, vol. 59, pp. 4820-4825, Dec 2011.
[10] E. Rubio-Martinez-Dueñas, de Jong van Coevorden C.M., and D. Caratelli, Supershaped complementary split-ring resonators, in 2017 IEEE AP-S Symposium on Antennas and Propagation and USNC-URSI Radio Science Meeting, 2017.
[11] J. D. Baena, J. Bonache, F. Martin, R. M. Sillero, F. Falcone, T. Lopetegi, M. A. G. Laso, J. Garcia-Garcia, I. Gil, M. F. Portillo, and M. Sorolla, Equivalent-circuit models for split-ring resonatorsand complementary split-ring resonators coupled to planar transmission lines,IEEE Transactions on Microwave Theory and Techniques, vol. 53, pp. 1451-1461, April 2005.
[12] C. Li, K. Liu, and F. Li, A microstriphighpass filter with complementary split ring resonators,Electronics Letters, vol. 43, pp. 35-36, 02 2007.
[13] V. Crnojevic-Bengin, V. Radonic, and B. Jokanovic, Fractal geometries of complementary split-ring resonators, IEEE Transactions on Microwave Theory and Techniques, vol. 56, pp. 2312{2321, Oct 2008.
[14] T. Li, G.-M. Wang, K. Lu, H.-X. Xu, Z.-H. Liao, and B. Zong, Novel bandpass filter based on CSRRs using Koch fractal curve,Progress in Electromagnetics Research Letters, vol. 28, pp. 121-128, 2012.
[15] R. Marques, F. Mesa, J. Martel, and F. Medina, Comparative analysis of edge- and broadside- coupled splitring resonators for metamaterial design - theory and experiments,IEEE Transactions on Antennas and Propagation, vol. 51, pp. 2572-2581, Oct 2003.
[16] J. D. Baena, R. Marques, F. Medina, and J. Martel, Artificial magnetic metamaterial design by using spiralresonators, Phys. Rev. B, vol. 69, p. 014402, Jan 2004.
[17] M. Palandoken and H. Henke, Fractal spiral resonator as magnetic metamaterial, in 2009 Applied Electromagnetics Conference (AEMC), pp. 1-4, Dec 2009.
[18] L. Youse_ and O. M. Ramahi, Artificial magnetic materials using fractal Hilbert curves,IEEE Transactions on Antennas and Propagation, vol. 58, pp. 2614-2622, Aug 2010.
[19] S. Zhou, Y.M. Xie, Q. Li, X. Huang, Fishnet metamaterials with double negative refractive index in the blue region of visible spectrum, Proc. of SPIE Vol 8806, 88062B, Metamaterials: Fundamentals and applications VI, Sept 11, 2013.

Advances on multi-scale MbD synthesis of WAIMs for advanced phased arrays

M Salucci and G Gottardi

ELEDIA Research Center (ELEDIA@UniTN - University of Trento)
Via Sommarive 9, I-38123 Trento, Italy
marco.salucci@unitn.it

Abstract. The most recent advances on the synthesis of wide-angle impedance matching (*WAIM*) devices for next-generation phased arrays are discussed. Towards this end, the *WAIM* design problem is formulated within the Material-by-Design (*MbD*) framework with the objective of minimizing the antenna input reflections caused by mutual coupling (*MC*) effects arising at the antenna aperture when steering the main beam in a wide angular region. Accordingly, the degrees-of-freedom (*DoF*s) are represented by the constituent materials of the synthesized structure and/or the *micro-scale* geometrical descriptors of the considered unit cells. Selected illustrative results will be shown in order to assess the effectiveness and the potentialities of leading-edge *MbD* solutions for the design of reliable and easy-to-implement *WAIM*s.

1. Introduction

Active electronically-scanned arrays (*AESA*s) represent nowadays a key-technology in many applicative domains ranging from radar to space communications [1]. Moreover, they will cover a fundamental role for the successful deployment of fifth-generation (*5G*) mobile communication systems [2]. However, the increasing demand of wide scanning angles and broadband functionalities are involving unprecedented design challenges in order to meet several conflicting requirements, which become particularly critical when aperture radiators (such as, for example, horns or truncated waveguides) are needed to allow high-power transmissions. Indeed, the scanning capabilities of such architectures are limited by the unavoidable mutual coupling (*MC*) effects arising between adjacent elements, causing non-negligible deviations of the reflection coefficient at the air-aperture interface when directing the main beam far from the broadside [3]-[8].

Several strategies can be adopted to reduce *MC* and enhance the array steering capabilities. Just to mention a few examples, non-uniform arrangements could be exploited to mitigate the inter-element interactions, with the main undesired drawback of decreasing the antenna efficiency [1]. Otherwise, low-profile multi-layer dielectric structures covering the array aperture, often indicated as wide-angle impedance matching (*WAIM*) devices can be designed to significantly reduce the amount of reflected power towards the feeding network in a wide range of steering angles and frequencies [3]-[8].

Within this context, artificially-engineered materials (such as, for instance, meta-materials/meta-surfaces) clearly represent one of the most appealing and competitive technologies to address the synthesis of *WAIM*s. Indeed, thanks to their capability of manipulating "at will" the propagation of electromagnetic (*EM*) waves, meta-materials are enabling the design of complex devices exhibiting unconventional features in many applicative fields [9]-[14], including (but not limited to) Luneburg lenses [11], mutual blockage reduction [13], and cloaking [9].

Content from this work may be used under the terms of the Creative Commons Attribution 3.0 licence. Any further distribution of this work must maintain attribution to the author(s) and the title of the work, journal citation and DOI.
Published under licence by IOP Publishing Ltd

IOP Conf. Series: Journal of Physics: Conf. Series **963** (2018) 012009 doi:10.1088/1742-6596/963/1/012009

The efficient realization of multi-layer and multi-frequency *WAIM*s has been recently proposed within the Material-by-Design (*MbD*) paradigm, an instance of the System-by-Design (*SbD*) framework [7], [8], [14], defined as the "*application-oriented synthesis of field-manipulating systems whose constituent electromagnetic properties are driven by the device functional requirements*". According to the *MbD* vision, the material electromagnetic (*EM*) properties [7] and/or the *micro-scale* structure of the elementary cell (when printed *WAIM*s are considered to simplify the realization process [8]) become the actual degrees-of-freedom (*DoF*s) to be properly tuned through effective and efficient optimization strategies, as briefly resumed in the following.

2. Multi-scale *MbD* approaches for the synthesis of *WAIM*s

Following the *MbD* paradigm [7], [8], [14], the problem of synthesizing a *WAIM* device providing the desired features is effectively addressed by decomposing the whole design process into elementary *functional blocks,* each solving rather "simple" tasks. More precisely, the solution is obtained by combining (*i*) a computationally-efficient *EM modeling block* for accurately modeling the reflections at the air-aperture interface, (*ii*) a *physical linkage block* devoted at assessing the fitting of all user-defined/application-driven constraints and objectives, and (*iii*) a *solution-space exploration block* for the effective search of the global optimum of a suitably defined cost function [7], [8].

To these functional blocks, a fourth *homogenization* one can be added to determine the equivalent permittivity/permeability tensors of each layer when considering the design of printed *WAIM*s over off-the-shelf dielectric substrates [8], [15]. As a matter of fact, the first *MbD* attempts to synthesize high-performance *WAIM*s involved the direct optimization of the constituent materials of each layer [3], [7]. However, despite the very promising results, those strategies rely on the availability (hopefully real in the near future) of manufacturing technologies able to reproduce arbitrary anisotropic permittivity and permeability distributions [7]. To overcome such a *feasibility* limit of the obtained solutions, *multi-scale MbD* approaches have been proposed [8], in which the *WAIM* is a printed meta-surface [16] rather than a set of homogeneous dielectric layers. In this case, the synthesis problem is re-formulated as the optimization of the *micro-scale* structure (i.e., the geometrical descriptors of the elementary cells), instead of the *macro-scale* equivalent permittivity/permeability tensors [8].

Concerning the actual implementation of each functional block, the modal analysis method [7] allows to derive an accurate but computationally efficient *EM modeling block* without recurring to time-consuming full-wave simulations. Regarding the *physical linkage block*, the *WAIM macro-scale* requirement is the minimization of the integral power reflection across all the considered steering angles (i.e., allowing a sufficiently large scan cone $\vartheta \in [\vartheta_{\min}, \vartheta_{\max}] \cup \varphi \in [\varphi_{\min}, \varphi_{\max}]$) and operative frequencies (i.e., enabling a proper matching over the user-defined frequency range $f \in [f_{\min}, f_{\max}]$). Such a requirement is typically mathematically translated into the following cost function [8]

$$\Phi(\mathbf{\Omega}) = \int_{f_{\min}}^{f_{\max}} \int_{\vartheta_{\min}}^{\vartheta_{\max}} \int_{\varphi_{\min}}^{\varphi_{\max}} \left|\Gamma(\vartheta, \varphi, f; \mathbf{\Omega})\right|^2 d\varphi d\vartheta df \tag{1}$$

where $\mathbf{\Omega}$ is the set of design *DoF*s (i.e., the layers thicknesses, the permittivity/permeability tensors [7] or the meta-surface unit-cell descriptors [8], etc...) and $\Gamma(\vartheta, \varphi, f; \mathbf{\Omega})$ is the voltage reflection coefficient at the planar aperture. Of course, additional terms could be added to (1), as well, to include multi-physics requirements depending on the applicative scenario (e.g., thermal, aerodynamic, etc...). As for the *solution-space exploration block*, remembering the *no-free-lunch* theorem on evolutionary optimization [17], [18] and given the highly non-linear/multi-modal nature of (1) and the real-nature of the *DoF*s, the Particle Swarm (*PS*) [7], [8], [18] represents a particularly suitable candidate to effectively and efficiently reach the global optimum. Finally, the *homogenization block* can be easily implemented through homogenization formulas linking the macro-scale *WAIM EM* properties (i.e., permittivity/permeability tensors) to its micro-scale features (i.e., the geometrical descriptors of the surface elementary cells) and the frequency [8], [15].

3. Conclusions

The most recent advances on the *MbD*-based synthesis of *WAIM* layers for enhancing the scanning capabilities of phased arrays have been discussed. The recently proposed methodologies allow unprecedented flexibility and effectiveness in realizing single-/multi-layer and single-/multi-frequency devices. Moreover, the development of *multi-scale MbD* approaches allowed the simplification of the manufacturing process through suitably designed microstrip printed surfaces. Further research is still needed to address the synthesis of complex unit-cells, as well as to generalize the existing strategies to any kind of phased array/elementary-radiator, and to conformal antennas. Furthermore, the integration of powerful interval-analysis (*IA*)-based techniques [19]-[24] in the *WAIM* design process is envisaged to predict and improve the robustness of the obtained layouts to several manufacturing uncertainties.

References

[1] Rocca P, Oliveri G, Mailloux RJ and Massa A 2016 Unconventional phased array architectures and design methodologies - A review *Proc. IEEE* **104** 544-560
[2] Oliveri G, Gottardi G, Robol F, Polo A, Poli L, Salucci M, Chuan M, Massagrande C, Vinetti P, Mattivi M, Lombardi R and Massa A 2017 Co-design of unconventional array architectures and antenna elements for *5G* base station *IEEE Trans. Antennas Propag.* **65** 6752-6767
[3] Sajuyigbe S, Ross M, Geren P, Cummer SA, Tanielian MH and Smith DR 2010 Wide angle impedance matching metamaterials for waveguide-fed phased-array antennas *IET Microw. Antennas Propag.* **4** 1063-1072
[4] Magill E and Wheeler HA 1966 Wide-angle impedance matching of a planar array antenna by a dielectric sheet *IEEE Trans. Antennas Propag.* **14** 49–53
[5] Araki K, Tanaka A and Matsumura E 2003 Wide scanning phased array antenna design in Ka band *IET Microw. Antennas Propag.* **150** 379–384
[6] Cameron TR and Eleftheriades GV 2015 Analysis and characterization of a wide-angle impedance matching metasurface for dipole phased arrays *IEEE Trans. Antennas Propag.* **63** 3928–3938
[7] Oliveri G, Viani F, Anselmi N and Massa A 2015 Synthesis of multilayer WAIM coatings for planar-phased arrays within the system-by-design framework *IEEE Trans. Antennas Propag.* **63** 2482-2496
[8] Oliveri G, Salucci M, Anselmi N and Massa A 2017 Multiscale system-by-design synthesis of printed WAIMs for waveguide array enhancement *IEEE J. Multiscale Multiphysics Computat. Techn.* **2** 84-96
[9] Eleftheriades GV and Engheta N, 2011 Metamaterials: fundamentals and applications in the microwave and optical regimes *Proc. IEEE* **99** 1618–1621
[10] Soric JC, Fleury R, Monti A, Toscano A, Bilotti F and Alu A 2014 Controlling scattering and absorption with metamaterial covers *IEEE Trans. Antennas Propag.* **62** 4220–4229
[11] Mateo-Segura C, Dyke A, Dyke H, Haq S and Hao Y 2014 Flat luneburg lens via transformation optics for directive antenna applications *IEEE Trans. Antennas Propag.* **62** 1945–1953
[12] Alu A, Bilotti F, Engheta N and Vegni L 2006 Metamaterial covers over a small aperture *IEEE Trans. Antennas Propag.* **54** 1632-1643
[13] Monti A, Soric J, Alu A, Bilotti F, Toscano A and Vegni L 2012 Overcoming mutual blockage between neighboring dipole antennas using a low-profile patterned metasurface *IEEE Antennas Wireless Propag. Lett.* **11** 1414-1417
[14] Nagar J, Campbell SD, Ren Q, Easum JA, Jenkins RP and Werner DH 2017 Multiobjective optimization-aided Metamaterials-by-Design with application to highly directive nanodevices *J. Multiscale Multiphysics Computat. Techn.* **2** 147-158
[15] Martini E, Sardi GM and Maci S 2014 Homogenization processes and retrieval of equivalent constitutive parameters for multisurface-metamaterials *IEEE Trans. Antennas Propag.* **62**, 2081-2092

[16] Minatti G, Martini E and Maci S 2017 Efficiency of metasurface antennas *IEEE Trans. Antennas Propag.* **65** 1532-1541
[17] Rocca P, Oliveri G Massa A 2011 Differential Evolution as applied to electromagnetics *IEEE Antennas Propag. Mag.* **53** 38-49
[18] Jin N and Rahmat-Samii Y 2007 Advances in particle swarm optimization for antenna designs: real-number, binary, single-objective and multiobjective implementations *IEEE Trans. Antennas Propag.* **55** 556–567
[19] Manica L, Anselmi N, Rocca P and Massa A 2013 Robust mask-constrained linear array synthesis through an interval-based particle swarm *IET Microw. Antennas Propag.* **7** 976-984
[20] Rocca P, Manica L, Anselmi N and Massa A 2013 Analysis of the pattern tolerances in linear arrays with arbitrary amplitude errors *IEEE Antennas Wireless Propag. Lett.*, **12** 639-642
[21] Anselmi N, Manica L, Rocca P and Massa A 2013 Tolerance analysis of antenna arrays through interval arithmetic *IEEE Trans. Antennas Propag.* **61** 5496-5507
[22] Rocca P, Anselmi N and Massa A 2014 Optimal synthesis of robust beamformer weights exploiting interval analysis and convex optimization *IEEE Trans. Antennas Propag.* **62** 3603-3612
[23] Rocca P, Anselmi N and Massa A 2014 Interval arithmetic for pattern tolerance analysis of parabolic reflectors *IEEE Trans. Antennas Propag.* **62** 4952-4960
[24] Poli L, Rocca P, Anselmi N and Massa A 2015 Dealing with uncertainties on phase weighting of linear antenna arrays by means of interval-based tolerance analysis *IEEE Trans. Antennas Propag.* **63** 3229-3234

Acknowledgment

This work benefited from the networking activities carried out within the FET Open CSA PROJECT NANOARCHITECTRONICS [grant Agreement number 737135] funded by the European Union in the framework of the Horizon 2020 – the Framework Programme for Research and Innovation (2014-2020).

IOP Conf. Series: Journal of Physics: Conf. Series **963** (2018) 012010 doi:10.1088/1742-6596/963/1/012010

Physically-insightful equivalent circuit models for electromagnetic periodic structures

F Mesa[1], R Rodríguez-Berral[1], F Medina[2]

[1] Dpt. Física Aplicada 1, Universidad de Sevilla, Sevilla, Spain

[2] Dpt. Electrnica & Electromagnetismo, Universidad de Sevilla, Sevilla, Spain

E-mail: mesa@us.es,rrberral@us.es,medina@us.es

Abstract.
In this presentation it will be discussed how to obtain analytical or quasi-analytical equivalent circuits to deal with periodic structures such as frequency selective surfaces and/or metasurfaces. Both the topology and the values of the involved elements of these circuits are obtained from a basic rationale to solve the corresponding integral equation. This procedure, besides providing a very efficient analysis/design tool, allows for a good physical insight into the operating mechanisms of the structure in contrast with the almost blind numerical scheme of commercial simulators.

1. Introduction

The study of periodic structures in Microwave Engineering has been a topic of intense research from early [1–3]. More recently the interest on this topic was boosted by the discovery of the so-called extraordinary optical transmission [4,5] and the further emergence of metasurfaces and related structures [6,7]. The study of these periodic or quasi-periodic structures finds a relevant precedent in the Frequency Selective Surfaces (FSSs), widely used to control the amplitude, phase and polarization of electromagnetic waves in antenna and scattering applications [8]. As is well known, a metasurface can be regarded as a special case of a FSS in which the electrical size of the constituent unit cell is small in comparison with the wavelength. In this case, the possibility of carrying out an "homogenization" procedure of the surface properties [9] conveniently allows for the treatment of the periodic surface as a sort of 2-D material [10]. Although metasurfaces and FSSs can be studied following different approaches, currently it is quite common to resort to electromagnetic commercial simulators, which, despite their high demands of computational resources, are now available in most research laboratories. Clearly, these commercial simulators have greatly benefited from the intense research carried out in the past decades on efficient numerical methods to deal with scattering problems [11].

2. Equivalent Circuit Approach

Another type of analysis of the periodic structures has been the Equivalent Circuit Approach (ECA), which aims at finding an equivalent network made up of transmission lines and lumped passive elements [12, 13]. This method has its roots in studies made in the fifties of the last century [14, 15]. Three different approaches can be distinguished in this line of action: i) obtaining the topology and values of the equivalent network from the *a priori* knowledge of

Content from this work may be used under the terms of the Creative Commons Attribution 3.0 licence. Any further distribution of this work must maintain attribution to the author(s) and the title of the work, journal citation and DOI.
Published under licence by IOP Publishing Ltd

the frequency response of the periodic structure [16]; ii) a general multi-mode analysis involving a "black-box" admittance matrix whose elements are derived from a rigorous numerical or semi-analytical analysis [17–19]; and iii) making use of some simplifying assumption (for instance, operation regime in the long-wavelength limit) that allows for a quasi-static analytical modeling of the structure's performance [9]. A clear drawback of the first family of methods is that the full dynamical behavior of the structure has to be known *a priori* in order to obtain a reduced-order model that works appropriately in the given frequency range of interest. Also, the network topology as well as the number and values of the involved elements are obtained from a purely numerical strategy. This general-purpose procedure provides little physical insight into the electromagnetic behavior of the structure, and the topology and elements of the equivalent network may vary depending on the frequency range of operation and/or the desired accuracy of the circuit model. The second family of methods is also a general tool that does not provide an explicit topology of the network, except in some limiting cases where it can be simplified. The third type of methods does provide a convenient physical insight but its range of application comes completely determined by the validity of the simplifying assumptions. Although there are numerous interesting situations covered by this last analytical approach, it fails for many other practical cases of interest. It is then the purpose of this talk to present an analytical approach that can be applied in a wide frequency range of operation but that preserves the advantages of the analytical derivation of an explicit topology of the ECA as well as closed-form expressions for the involved lumped elements and transmission lines. Similar relevant attempts in this direction have been reported in the past [20] and some examples of more recent effort of the authors of this presentation to extend the validity and applicability of this type of ECA can be found in [21, 22].

Acknowledgments

This work has been supported by the Spanish Ministerio de Economía y Competitividad with European Union FEDER funds (project TEC2013-41913-P) and by the Spanish Junta de Andalucía (project P12-TIC-1435).

References

[1] G. V. Trentini, "Partially reflecting sheet arrays," *IRE Trans. Antennas Propag.*, vol. 4, pp. 666-671, 1956.
[2] R. Ulrich, "Far-infrared properties of metallic mesh and its complementary structure," *Infrared Phys.*, vol. 7, pp. 37-55, 1967.
[3] A. Hessel, A. A. Oliner, M. H. Chen, and R. C. M. Li, "Propagation in periodically loaded waveguides with higher symmetries," *Proc. IEEE*, vol. 61, no. 2, pp. 183-195, 1973.
[4] T. W. Ebbesen, H. J. Lezec, H. F. Ghaemi, T. Thio, and P. A. Wolff, "Extraordinary optical transmission through sub-wavelength hole arrays," *Nature*, vol. 391, pp. 667-669, Feb. 1998.
[5] F. J. García-de-Abajo, " Colloquium: light scattering by particle and hole arrays," *Rev. Mod. Phys.*, vol. 79, pp. 1267-1290, Oct.-Dec. 2007.
[6] D. F. Sievenpiper, *et al*, "Two-dimensional beam steering using an electrically tunable impedance surface," *IEEE Trans. Antennas Propag.*, vol. 51, no. 10, pp. 2713-2722, Oct. 2003.
[7] S. Maci, G. Minatti, M. Casaletti, and M. Bosiljevac, "Metasurfing: addressing waves on impenetrable metasurfaces," *IEEE Antennas Wirel. Propag. Lett.*, vol. 10, pp. 1499-1502, 2011.
[8] B. Munk, *Frequency Selective Surfaces: Theory and Design*, Ed. Wiley, 2000.
[9] O. Luukkonen *et al*, "Simple and accurate analytical model of planar grids and high-impedance surfaces comprising metal strips or patches," *IEEE Trans. Antennas Propag.*, vol. 56, no. 6, pp. 1624-1632, June 2008.
[10] C. L. Holloway, E. F. Kuester, and A. Dienstfrey, "Characterizing metasurfaces/metafilms: The connection between surface susceptibilities and effective material properties," *IEEE Antennas Wirel. Propag. Lett.*, vol. 10, pp. 1507-1511, 2011.
[11] A. F. Peterson, S. L. Ray, and R. Mittra, *Computational Methods for Electromagnetics*. IEEE Press, 1998.
[12] I. Palocz and A. A. Oliner, "Equivalent network of a multimode planar grating," *IEEE Trans. Microw. Theory Techn.*, vol. 18, pp. 244–252, May 1970.
[13] C. K. Lee and R.J. Langley, "Equivalent-circuit models for frequency-selective surfaces at oblique angles of incidence," *IEE Proceedings*, vol. 132, Pt. H, no. 6, pp. 395-399, Oct. 1985.

[14] C. G. Montgomery, R. H. Dicke, and E. M. Purcell. *Principles of Microwave Circuits*, MIT Radiation Laboratory Series, vol. 8, McGraw-Hill, New York, 1948.

[15] N. Marcuvitz, *Waveguide Handbook*, MIT Radiation Laboratory Series, vol. 10, McGraw-Hill, New York 1951. New Ed., IEE Publishing/Peregrinus, 1986.

[16] S. Maci, M. Caiazzo, A. Cucini, and M. Casaletti, "A pole-zero matching method for EBG surfaces composed of a dipole FSS printed on a grounded dielectric slab," *IEEE Trans. Antennas Propag.*, vol. 53, no. 1, pp. 70-81, Jan. 2005.

[17] M. Guglielmi and A. A. Oliner, "Multimode network description of a planar periodic metal-strip grating at a dielectric interface - Part I: Rigorous network formulations," *IEEE Trans. Microw. Theory Techn.*, vol. 37, no. 3, pp. 535–541, March 1989.

[18] G. Conciauro, M. Guglielmi, and R. Sorrentino, *Advanced Modal Analysis*, Ed. Wiley, 1999.

[19] S. Monni, G. Gerini, A. Neto, and A. G. Tijhuis, "Multi-mode equivalent networks for the design and analysis of frequency selective surfaces," *IEEE Trans. Antennas Propag.*, vol. 55, no. 10, pp. 2824-2835, Oct. 2007.

[20] R. Dubrovka, J. Vazquez, C. Parini, and D. Moore, "Equivalent circuit method for Analysis and synthesis of frequency selective surfaces," *IEE Proc. pt. H, Microw. Antennas Propag.*, vol. 153, no. 3, pp. 213-220, March 2006.

[21] R. Rodríguez-Berral, F. Mesa, and F. Medina", "Analytical multimodal network approach for 2-D arrays of planar patches/apertures embedded in a layered medium," *IEEE Trans. Antennas Propag.*, vol. 63, pp. 1969-1984, 2015.

[22] V. Torres, F. Mesa, M. Navarro-Cía, R. Rodríguez-Berral, M. Beruete, and F. Medina, "Accurate circuit modeling of fishnet structures for negative-index-medium applications," *IEEE Trans. Microw. Theory Techn.*, vol. 64, pp. 15-26, Jan. 2016.

IOP Conf. Series: Journal of Physics: Conf. Series **963** (2018) 012011 doi:10.1088/1742-6596/963/1/012011

Recent Advances and Current Trends in Metamaterial-by-Design

N Anselmi and G Gottardi

[1] ELEDIA Research Center (ELEDIA@UniTN – University of Trento), Via Sommarive 9, 38123 Trento – Italy.
{nicola.anselmi, giorgio.gottardi}@eledia.org

Abstract. Thanks to their potential applications in several engineering areas, metamaterials gained much of attentions among different research communities, leading to the development of several analysis and synthesis tools. In this context, the metamaterial-by-design (*MbD*) paradigm has been recently introduced as a powerful tool for the design of complex metamaterials-based structures. In this work a review of the state-of-art, as well as the recent advancements of *MbD*-based methods are presented.

1. Introduction and Metamaterial-by-Design Paradigm

In the last years artificial metamaterials attract the attention of researchers from different engineering areas, thanks to the several kinds of applications that effectively exploit the promising electromagnetic (*EM*) features of this class of artificial materials [1][2]. Anyway, the design of metamaterials when integrated into complex *EM* systems is a challenging task. The main issues are related to the time consuming optimizations, as well as non-optimal integrations of subparts of the complex system during the optimization of the metamaterial, leading to sub-optimal solutions [3]. For this reasons, the *Metamaterial-by-Design* (*MbD*) paradigm has been introduced, affording a new synthesis strategy based on a modular representation of the design process [3]. By sub-dividing the whole synthesis/analysis procedure, into several interconnected functional blocks, an *MbD* synthesis procedure allows to drive the design of the metamaterial towards the fulfilment of the system functional requirements. More in detail this novel design approach permits to select the most suitable methodology within each functional block in order to reach optimal trade-offs depending on the design constraints/requirements, and to combine together the advantages of different existing and novel techniques within the same design tool. Four main blocks compose the *MbD* synthesis loop: i) a global optimizer for the optimization of the geometric descriptors and dielectric properties of the elementary metamaterial cell; ii) an homogenization functional block, devoted to the computation of the macro equivalent homogenized dielectric parameters; iii) the forward solver tool, for a fast computation of the *EM* behaviour of the metamaterial device; iv) a physical linkage analysis module aimed at verify if the current solution fits the design goals and constraints, giving feedback to the optimization block. According to this synthesis scheme, different *EM* synthesis problems have been effectively addressed, starting from the design of radome-coated arrays for array miniaturization [4]-[6] integrating a quasi-conformal transformation optics (*QCTO*) method and a source inversion technique [4][5], and exploiting spline basis functions in order to reduce the number of degrees-of-freedom handled by the global optimizer [6]. These flexible design tools allow to reduce the number of radiators, having radiation properties that are close to the corresponding wider array, and to obtain trade-off solutions between the complexity/feasibility of the synthesized radome material and the array miniaturization.

Content from this work may be used under the terms of the Creative Commons Attribution 3.0 licence. Any further distribution of this work must maintain attribution to the author(s) and the title of the work, journal citation and DOI.
Published under licence by IOP Publishing Ltd

In [7] the design of electromagnetic surface wave devices combining transformation optics and modal analysis has been recently proposed, while in [8][9] the synthesis of modulated metasurfaces with polarization control [8] and with multi-beam functionalities [9] have been presented, as well as metamaterial-based design methodologies for broadband aperture antennas [10] and dual-circularly polarized antenna based on the field transformation [11]. In [12][13] the problem of adding new features or to enhance the performances of linear phased arrays, without increasing the complexity of the feeding network, has been addressed exploiting the *MbD* paradigm, while in [14] the radiation efficiency of planar phased arrays has been enhanced by using layers of wide-angle impedance matching (*WAIM*) printed metasurfaces, by means of a multi-scale *MbD*-based methodology. Innovative design approaches for the synthesis of sectoral/conical horn antennas with advanced polarization functionalities [15] and compact layouts [16] have been recently proposed, exploiting metamaterial properties. Recently, multi-objective optimization strategies for the synthesis of highly directive nanodevices [17] and three dimensional frequency selective surfaces [18] have been proposed within the *MbD* framework. The presented instances of the *MbD* paradigm show that it represents a powerful approach for the design of future metamaterial-based complex systems, with promising potentialities on many other possible applicative contexts.

Acknowledgments
This work benefited from the networking activities carried out within the FET Open CSA PROJECT NANOARCHITECTRONICS [grant Agreement number 737135] funded by the European Union in the framework of the Horizon 2020 – the Framework Programme for Research and Innovation (2014-2020).

References

[1] Oliveri G, Werner D. H. and Massa A 2015 Reconfigurable electromagnetics through metamaterials - A review *Proc. IEEE* **103** 1034-56
[2] Tretyakov S, Urbas A and Zheludev N 2017 The century of metamaterials *Journal of Optics* **19**
[3] Massa A and Oliveri G 2016 Metamaterial-by-design: theory, methods, and applications to communications and sensing – Editorial *EPJ Appl. Metamat.* **3** 1-3
[4] Oliveri G, Bekele E T, Salucci M and Massa A 2015 Array miniaturization through QCTO-SI metamaterial radomes *IEEE Trans. Antennas Propag.* **63** 3465-76
[5] Salucci M, Oliveri G, Ahmadi H and Massa A 2016 Conformal transformation of linear arrays through QCTO-based design tools *Proc. IEEE AP-S Int. Symp. and USNC-URSI Radio Science Meeting* (Fajardo Puerto Rico) 1083-84
[6] Tenuti L, Oliveri G, Viani F and Massa A 2015 Spline-enhanced synthesis of metamaterial lenses for linear array miniaturization by the SbD-QCTO *Proc. IEEE AP-S Int. Symp. and USNC-URSI Radio Science Meeting* (Vancouver Canada) 951-2
[7] La Spada L, Haq S and Hao Y 2016 Modeling and design for electromagnetic surface wave devices *Radio Sci.* **52** 1049-1057
[8] Minatti G, Caminita F, Martini E, Sabbadini M and Maci S 2016 Synthesis of modulated-metasurface antennas with amplitude, phase, and polarization control *IEEE Trans. Antennas Propag.* **64** 3907-19
[9] González-Ovejero D, Minatti G, Chattopadhyay G and Maci S 2017 Multibeam by metasurface antennas *IEEE Trans. Antennas Propag.* **65** 2923-30
[10] Ramaccia D, Barbuto M, Monti A, Verrengia A, Trotta F, Muha D, Hrabar S, Bilotti F and Toscano A 2016 Exploiting intrinsic dispersion of metamaterials for designing broadband aperture antennas: theory and experimental verification *IEEE Trans. Antennas Propag.* **64** 1141-46
[11] Shi H, Giddens H and Hao Y 2017 Dual-circularly polarized patch antenna using field transformation medium *IEEE Antennas Wireless Propag. Lett.* **16** 2869-72
[12] Salucci M, Tenuti L, Bekele E T and Oliveri G 2016 Enhancement of linear arrays through

MbD metamaterial coatings *Proc. IEEE AP-S Int. Symp. and USNC-URSI Radio Science Meeting* (Fajardo Puerto Rico) 1089-90

[13] Salucci M, Oliveri G, Anselmi N, Gottardi G and Massa A 2017 Performance enhancement of linear active electronically-scanned arrays by means of MbD-synthesized metalenses *J. Electromagnet. Wave.* 1-29

[14] Oliveri G, Salucci M, Anselmi N and Massa A 2017 Multiscale system-by-design synthesis of printed WAIMs for waveguide array enhancement *IEEE J. Multiscale Multiphysics Computat. Techn.* **2** 84-96

[15] Ramaccia D, Sounas D L, Alù A, Bilotti F and Toscano A 2015 Nonreciprocal horn antennas using angular momentum-biased metamaterial inclusions *IEEE Trans. Antennas Propag.* **63** 5593-5600

[16] Oliveri G, Bekele E T, Salucci M and Massa A 2016 Transformation electromagnetics miniaturization of sectoral and conical horn antennas *IEEE Trans. Antennas Propag.* **64** 1508-13

[17] Nagar J, Campbell S D, Ren Q, Easum J A, Jenkins R P and Werner D H 2017 Multiobjective optimization-aided metamaterials-by-design with application to highly directive nanodevices *IEEE J. Multiscale Multiphysics Computat. Techn.* **2** 147-158

[18] Zhu D Z, Werner P L and Werner D H 2017 Design and optimization of 3-D frequency-selective surfaces based on a multiobjective lazy ant colony optimization algorithm *IEEE Trans. Antennas Propag.* **65** 7137-49

Application of metamaterial concepts to sensors and chipless RFID

F. Martín, C. Herrojo, P. Vélez, L. Su, J. Mata-Contreras, F. Paredes

CIMITEC, Departament d'Enginyeria Electrònica, Universitat Autònoma de Barcelona, 08193 Bellaterra, Barcelona, Spain

E-mail: Ferran.Martin@uab.es

Abstract. Several strategies for the implementation of microwave sensors based on the use of metamaterial-inspired resonators are pointed out, and examples of applications, including sensors for dielectric characterization and sensors for the measurement of spatial variables, are provided. It will be also shown that novel microwave encoders for chipless RFID systems with very high data capacity can be implemented. The fields of applications of the devices discussed in this talk include dielectric characterization of solids and liquids, angular velocity sensors for space applications, and near-field chipless RFID systems for secure paper applications, among others.

1. Introduction

In this review paper, it is shown that transmission lines loaded with electrically small metamaterial-inspired resonators can be applied in many diverse scenarios, including sensing of spatial variables, dielectric characterization of solids/liquids, and chipless radiofrequency identification (chipless-RFID). Transmission line metamaterials based on electrically small resonators (particularly split ring resonators –SRRs) were first proposed in [1], where it was shown that, by loading a coplanar waveguide (CPW) with SRRs and shunt inductive strips, left handed wave propagation was possible. On the basis of impedance and dispersion engineering, transmission line metamaterials based on SRRs and other resonant elements (e.g., complementary split ring resonators –CSRRs [2]) have been applied to the design of multiple devices exhibiting improved performance or novel functionalities (filters [3], dual-band components [4,5], enhanced bandwidth components [6], leaky-wave antennas [7], etc.).

In other applications of transmission lines loaded with SRRs or CSRRs (or related metamaterial particles), the resonance phenomenon, rather than dispersion and impedance engineering is exploited [8]. For example, resonator-loaded lines can be applied to the implementation of notch filters. Moreover, as long as the resonance frequency in such lines (transmission zero) is influenced by the region surrounding the resonant element and by the relative position/orientation between the line and the resonator, sensing of material properties [9-34] and spatial variables [35-47] is possible. It is also possible to use chains of resonant elements printed on a dielectric substrate for the implementation of microwave encoders (tags) useful for applications in chipless-RFID. In such RFID systems, tag reading is achieved by means of a dedicated transmission line, through near-field coupling to the tag [48-52].

This paper is focused on reviewing three applications of transmission lines loaded with metamaterial resonators: (i) sensors for angular displacement and velocity measurements, (ii) sensors

Content from this work may be used under the terms of the Creative Commons Attribution 3.0 licence. Any further distribution of this work must maintain attribution to the author(s) and the title of the work, journal citation and DOI.
Published under licence by IOP Publishing Ltd

for dielectric characterization, and (iii) microwave encoders for chipless RFID. The principle of operation is pointed out for each case, and an illustrative example is provided.

2. Angular displacement and velocity sensors

The type of angular displacement and velocity sensor reported here is based on a transmission line (CPW) loaded with an electrical-LC (ELC) resonator. Such sensors were first reported in [40], and then implemented in microstrip technology in [41], and by using S-shaped SRRs in [44]. The stator is a CPW circularly shaped, whereas the rotor contains a circular ELC resonator axially attached to it (Fig. 1). The ELC of the rotor is placed on top of the CPW in close proximity and parallel to it, so that coupling between the line and the resonator is possible. The working principle is based on coupling modulation by rotation. Namely, the ELC is a bi-symmetric particle, exhibiting an electric wall and a magnetic wall (orthogonally oriented) at the fundamental resonance. If the magnetic wall of the particle is aligned with the line axis (also a magnetic wall), coupling arises, and the transmission coefficient exhibits a notch with significant depth. Conversely, by rotating the particle 90°, i.e., by aligning the electric wall of the ELC with the line axis, coupling is prevented, and the line exhibits total transmission. Obviously, between these two (extreme) situations, the coupling (and hence the notch depth) is modulated by rotation and the relative angle between the line and the ELC can be determined (Fig. 2).

(a) (b)

Fig. 1. Photograph of the stator (a) and rotor (b).

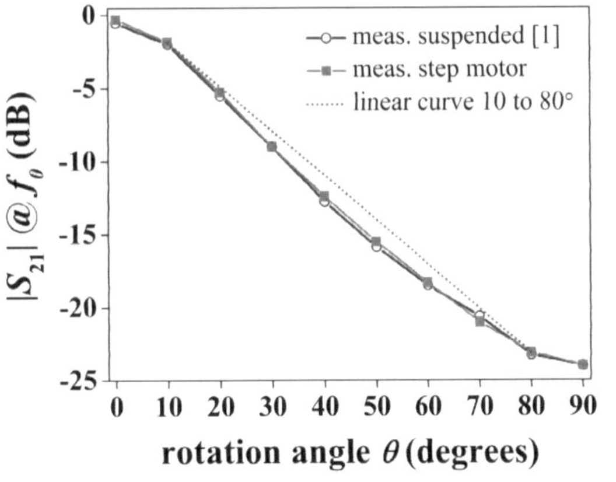

Fig. 2. Transfer function of the notch magnitude at the notch frequency in the transmission coefficient versus the rotation angle. Reprinted with permission from [40].

For the measurement of the angular velocity, a harmonic (carrier) signal is injected to the input port of the stator. Through rotor motion, line to resonator coupling is modulated with the result of an amplitude modulated (AM) signal at the output. From the distance between adjacent maxima of the envelope function, the angular velocity can be inferred, provided maximum transmission occurs twice per cycle. The envelope function can be obtained by means of an envelope detector (implemented by a diode and low-pass filter) preceded by an isolator, in order to prevent unwanted reflections from the diode. Figure 3 shows the principle of operation, the experimental set-up is depicted in Fig. 4, whereas Fig. 5 shows the envelope function obtained by an oscilloscope (corresponding to an angular velocity of 3.000 rpm).

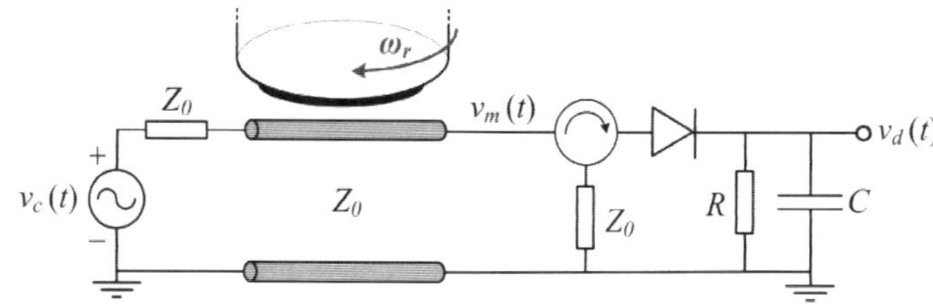

Fig. 3. Schematic of the angular velocity system.

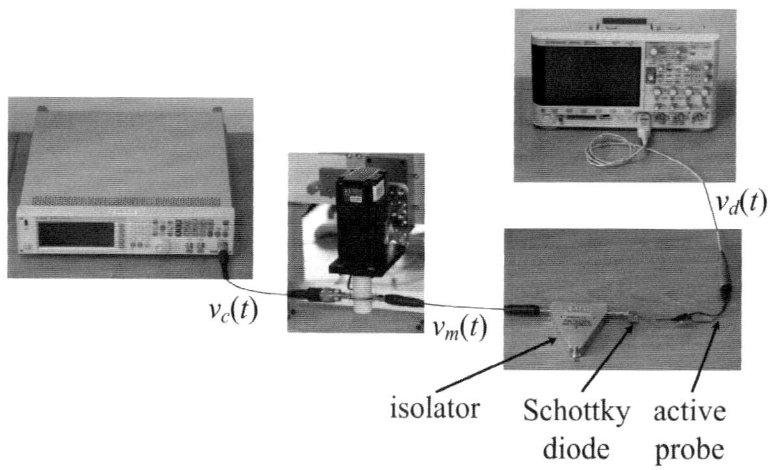

Fig. 4. Photographs of the experimental implementation. Reprinted with permission from [40].

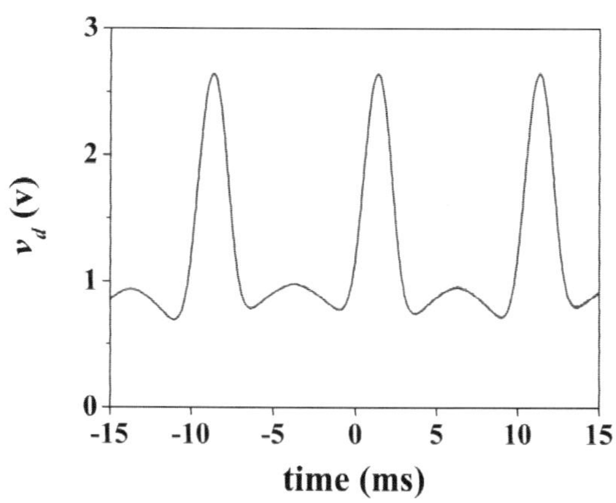

Fig. 5. Measured envelope signal for an angular velocity of 3.000 rpm. Reprinted with permission from [40].

3. Sensors for dielectric characterization

There are many types of sensors for dielectric characterization based on resonator-loaded lines. Here we report sensors based on frequency splitting [27,29-32]. By symmetrically loading a line with two identical resonators coupled to it, a single notch at the fundamental resonance of the resonant elements arises. However, by truncating symmetry, e.g., by means of an asymmetric dielectric loading, two notches arise, and the frequency separation between them as well as the difference in the notch depths depends on the level of asymmetry. Therefore, the determination of material properties, specifically the permittivity, of a sample under test (SUT), as compared to a reference sample, can be inferred.

In this paper we report a sensor based on a splitter/combiner configuration [31,32,34], where the splitter/combiner is loaded with a pair of SRRs (Fig. 6) [34]. The sensor is equipped with a pair of microfluidic channels, in order to be able to measure the complex dielectric constant of liquids, particularly, mixtures of deionized (DI) water and ethanol, considering DI water as reference sample (with well known dielectric properties). Fig. 7 depicts the response of the structure for various combination of DI water and ethanol for the SUT, and Fig. 8 shows the variation of the real and imaginary parts of the dielectric constant, inferred from the response of Fig. 7 according to the method detailed in [34].

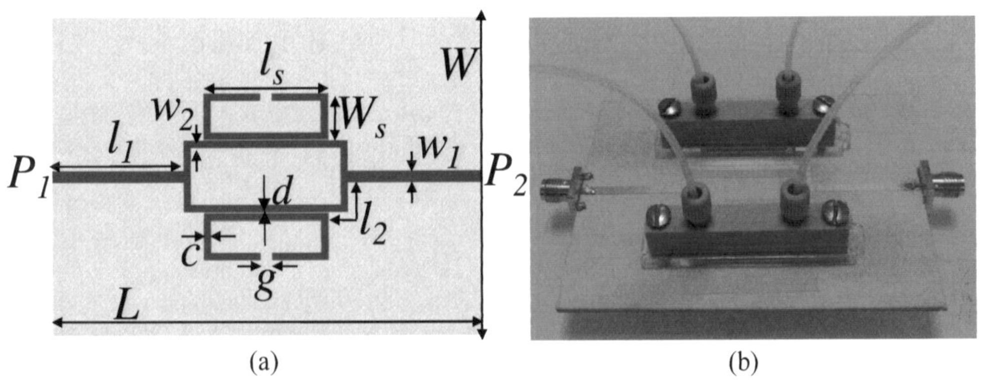

Fig. 6. Topology of the splitter/combiner configuration and relevant dimensions (a) and photograph of the structure including the microfluidic channels for sensing (b). L = 86 mm, W = 62 mm, inverter dimensions are l_1 = 27 mm and w_1 = 2.22 mm, SRR dimensions are l_s = 25 mm, W_s = 9 mm, c = 1.4 mm, g = 2.4 mm, the slot separation between the lines and the SRRs is d = 0.2 mm, and the dimensions of the transmission lines sections between the T-junctions and the SRRs are l_2 = 9.21 mm, w_2 = 1.34 mm. The ground plane is depicted in grey. Reprinted with permission from [34].

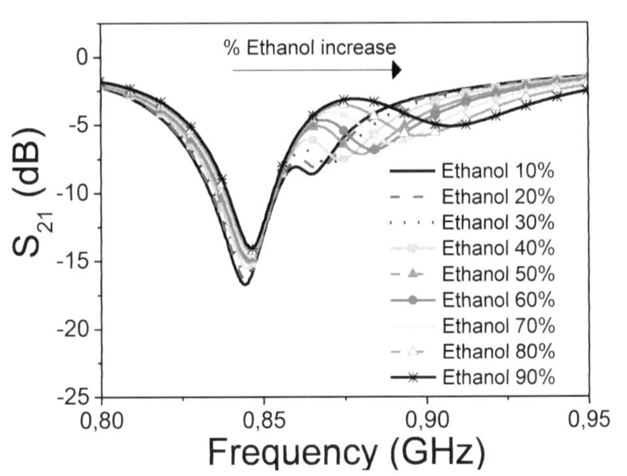

Fig. 7. Measured transmission coefficient (magnitude) for different mixtures of DI water and ethanol (LUT), considering DI water as reference liquid. Reprinted with permission from [34].

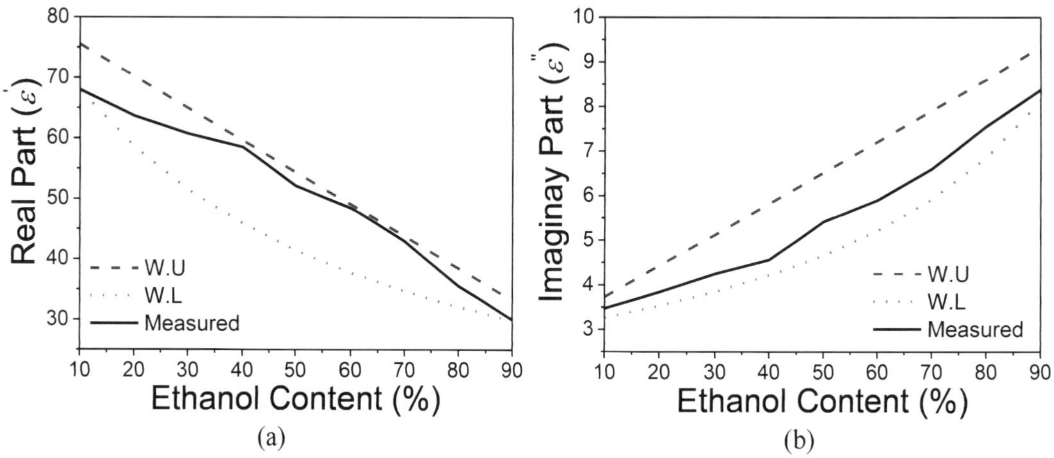

Fig. 8. Extracted value for the real (a) and imaginary (b) parts of the complex dielectric constant in mixtures of DI water/ethanol. The static Weiner model [53] (upper and lower limits) is also included for comparison purposes Reprinted with permission from [34].

4. Microwave encoders for chipless RFID

Typically, RFID systems, used for identification and tracking, are based on the use of tags equipped with chips, where the information relative to the object or item is stored. Chipped tags are relatively expensive for many applications involving low-cost items. An alternative to RFID systems based on chipped tags is chipless-RFID [54]. In such systems, the chip is replaced with a planar passive encoder, which contains the identification (ID) code. There are many types of encoders, working in time domain [55-63], frequency domain [48-52,64-80], or even hybrid encoders [77-80], the latter exploiting several domains simultaneously. The main drawback of chipless-RFID systems is the limited data storage capability of the tags (encoders) as well as encoder size. Here we report a chipless-RFID system, where the number of bits is only limited by tag size.

The approach, first reported in [51], and subsequently improved in various works [52,81,82], is based on near-field coupling between the reader and the tag and sequential bit reading. The tag is a chain of identical resonant elements etched or printed at equidistant and predefined positions on a dielectric substrate. The presence or absence of resonant elements in such positions determines the ID code. An alternative to tag absence in a certain position is resonator detuning by short-circuiting it, or by any other mean to drastically shift the resonance frequency (i.e., laser ablation, etc.). Tag reading is achieved by means of a transmission line fed with a harmonic signal tuned to the resonance frequency of the tag resonators. In a reading operation, the tag should be displaced above the transmission line of the reader, in close proximity to it, so that the resonant elements sequentially cross the line axis. By this means each time a resonator crosses the line, line-to-resonator coupling prevents the harmonic signal to be transmitted, thus generating an amplitude modulated (AM) signal that contains the ID code of the tag. As compared to other chipless-RFID systems based on multiple resonators, each one tuned to a different frequency (spectral signature barcodes), the proposed tags exhibit much larger data capacity, only limited by tag size. This is due to the fact that the feeding signal is a single tone signal, contrary to the multi-frequency (sweeping) signal required to read spectral signature barcodes, forcing the tag to exhibit a limited bandwidth and hence number of bits.

Figure 9 sketches the proposed chipless-RFID system, whereas Fig. 10 depicts the transmission line of the reader, as well as a 10-bit tag with all resonators (S-SRRs) present at the predefined positions (corresponding to the code '1111111111'). It is interesting to mention that the line is loaded with an S-SRR as well (rotated 180° with regard to the S-SRRs of the tag). This is done in order to avoid inter-resonator coupling in the tag as well as coupling between the line and multiple tag resonators, as discussed in [51]. Figure 11 shows the response of various tags with the indicated code, pointing out the validity of the proposed system. This near-field chipless-RFID system with sequential bit reading is especially useful in security paper applications, e.g. to avoid unauthorized copies, counterfeiting, etc. of corporate and official documents, ballots, certificates,... Note also that the reported tags have only 10 bits, but this number can be increased at convenience, with the unique limitation of tag size.

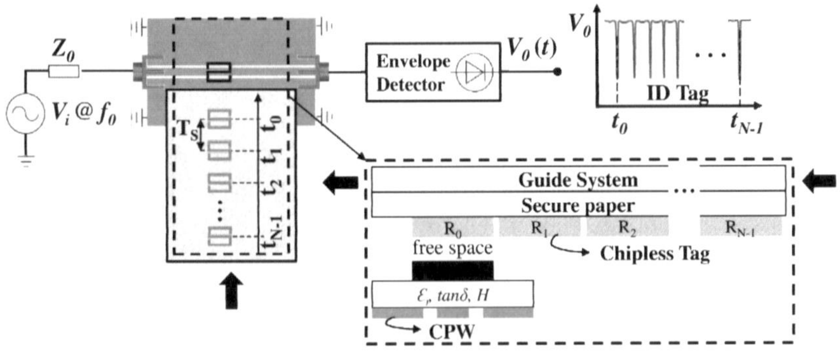

Fig. 9. Illustration of the working principle of the proposed chipless RFID system. The cross sectional view is depicted in the inset. The black arrows indicate tag motion in a reading operation. Reprinted from [51] with permission.

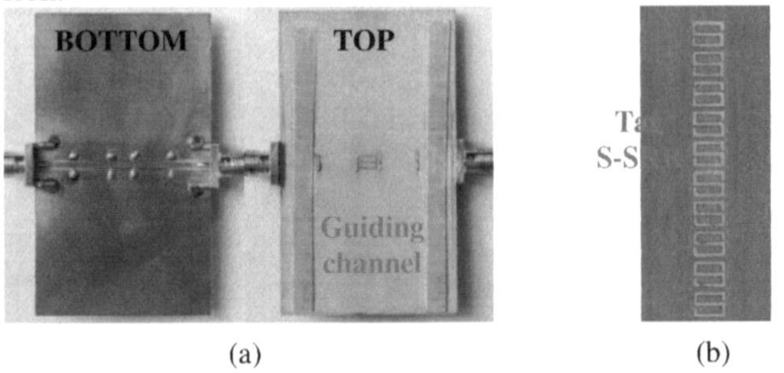

Fig. 10. Transmission line reader consisting of the S-SRR-loaded CPW (a) and photograph of the fabricated encoder (b). CPW dimensions are (in mm) $W = 1.2$ and $G = 0.48$; S-SRR dimensions are (in mm): length $l_1 = 3.8$, width $l_2 = 2.96$, strip width $c_0 = 0.4$ and gap space $s = 0.2$. Reprinted from [51] with permission.

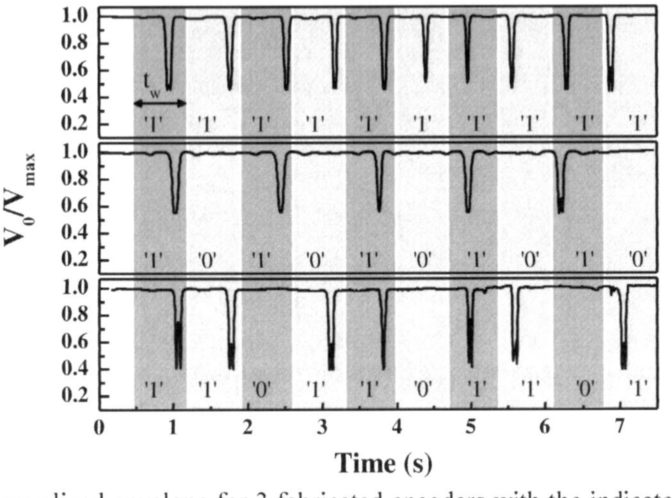

Fig. 11. Measured normalized envelope for 3 fabricated encoders with the indicated codes. Reprinted from [51] with permission.

5. Conclusions

In conclusion, it has been shown that transmission lines loaded with metamaterial resonant elements are useful for many diverse purposes, including sensing of material properties, sensing of spatial variables and velocities, and implementation of high data capacity chipless-RFID systems based on near-field coupling.

Acknowledgments

This work was supported by MINECO-Spain (projects TEC2013-40600-R, TEC2016-75650-R and RTC-2014-2550-7), Generalitat de Catalunya (project 2014SGR-157), Institució Catalana de Recerca i Estudis Avançats (who awarded F. Martín), and by FEDER funds. C. Herrojo acknowledges MINECO for supporting his research activity through the FPI grant BES-2014-068164.

References

[1] F. Martín, F. Falcone, J. Bonache, R. Marqués and M. Sorolla, "Split ring resonator based left handed coplanar waveguide", Appl. Phys. Lett., vol. 83, pp. 4652-4654, Dec. 2003.

[2] F. Falcone, T. Lopetegi, J.D. Baena, R. Marqués, F. Martín and M. Sorolla, "Effective negative-stop-band microstrip lines based on complementary split ring resonators", IEEE Microwave and Wireless Components Letters, vol. 14, pp. 280-282, Jun. 2004.

[3] J. Bonache, I. Gil, J. García-García, F. Martín, "Novel Microstrip Band Pass Filters Based on Complementary Split Rings Resonators", IEEE Transactions on Microwave Theory and Techniques, vol. 54, pp. 265-271, Jan. 2006.

[4] J. Bonache, G. Sisó, M. Gil, A. Iniesta, J. García-Rincón and F. Martín, "Application of composite right/left handed (CRLH) transmission lines based on complementary split ring resonators (CSRRs) to the design of dual band microwave components", IEEE Microwave and Wireless Components Letters, vol. 18, pp. 524-526, August 2008.

[5] M. Durán-Sindreu, A. Vélez, F. Aznar, G. Sisó, J. Bonache and F. Martín, "Application of Open Split Ring Resonators and Open Complementary Split Ring Resonators to the Synthesis of Artificial Transmission Lines and Microwave Passive Components", IEEE Trans. Microwave Theory and Techniques, vol. 57, pp. 3395-3403, Dec. 2009.

[6] G. Sisó, J. Bonache, M. Gil and F. Martín, "Application of resonant-type metamaterial transmission lines to the design of enhanced bandwidth components with compact dimensions", Microwave and Optical Technology Letters, vol. 50, pp. 127-134, January 2008.

[7] G. Zamora, S. Zuffanelli, F. Paredes, F. Javier Herraiz-Martínez, F. Martín, and J. Bonache, "Fundamental mode leaky-wave-antenna (LWA) using slot line and split-ring-resonator (SRR)-based metamaterials", IEEE Antennas and Wireless Propagation Letters, vol. 12, pp. 1424-1427, 2013.

[8] F. Martín, *Artificial Transmission Lines for RF and Microwave Applications*, John Wiley, Hoboken, USA, 2015.

[9] M. S. Boybay and O. M. Ramahi, "Material characterization using complementary split-ring resonators," *IEEE Trans. Instrum. Meas.*, vol. 61, no. 11, pp. 3039–3046, Nov. 2012.

[10] C.-S. Lee and C.-L. Yang, "Complementary split-ring resonators for measuring dielectric constants and loss tangents," *IEEE Microw. Wireless Compon. Lett.*, vol. 24, no. 8, pp. 563–565, Aug. 2014.

[11] C.-L. Yang, C.-S. Lee, K.-W. Chen, and K.-Z. Chen, "Noncontact measurement of complex permittivity and thickness by using planar resonators," *IEEE Trans. Microw. Theory Techn.*, vol. 64, no.1, pp. 247–257, Jan. 2016.

[12] M. Puentes, C. Weiß, M. Schüßler, and R. Jakoby, "Sensor array based on split ring resonators for analysis of organic tissues," in *IEEE MTT-S Int. Microw. Symp.*, Baltimore, MD, USA,

Jun. 2011, pp. 1–4.

[13] M. Puentes, Planar Metamaterial Based Microwave Sensor Arrays for Biomedical Analysis and Treatment. Springer, Heidelberg, Germany, 2014.

[14] T. Chretiennot, D. Dubuc, and K. Grenier, "A microwave and microfluidic planar resonator for efficient and accurate complex permittivity characterization of aqueous solutions," *IEEE Trans. Microw. Theory Techn.*, vol. 61, no. 2, pp. 972–978, Feb. 2013.

[15] A. Abduljabar, D. Rowe, A. Porch, and D. Barrow, "Novel microwave microfluidic sensor using a microstrip split-ring resonator," *IEEE Trans. Microw. Thory Techn.*, vol. 62, no. 3, pp. 679–688, Mar. 2014.

[16] A. Ebrahimi, W. Withayachumnankul, S. Al-Sarawi, D. Abbott, "High-sensitivity metamaterial-inspired sensor for microfluidic dielectric characterization," *IEEE Sensors J.*, vol. 14, no. 5, pp. 1345–1351, May 2014.

[17] W. Withayachumnankul, K. Jaruwongrungsee, A. Tuantranont, C. Fumeaux, and D. Abbott, "Metamaterial-based microfluidic sensor for dielectric characterization," *Sensor Actuat. A Phys.*, vol. 189, pp. 233–237, Jan. 2013.

[18] T. Chretiennot, D. Dubuc, and K. Grenier, "Optimized Electromagnetic Interaction Microwave Resonator/Microfluidic Channel for Enhanced Liquid Bio-Sensor," in *Europ. Microw. Conf.*, pp. 464–467, Dec. 2013.

[19] A. Salim and S. Lim, "Complementary Split-Ring Resonator-Loaded Microfluidic Ethanol Chemical Sensor," *Sensors*, vol. 16, pp. 1-13, 2016.

[20] H.-J. Lee and J.-G. Yook, "Biosensing using split-ring resonators at microwave regime," *App. Phys. Lett.*, vol. 92, no. 25, p. 254103, 2008.

[21] E. Ekmekci and G. Turhan-Sayan, "Multi-functional metamaterial sensor based on a broad-side coupled SRR topology with a multi-layer substrate," *App. Phys. A* , vol. 110, no. 1, pp. 189–197, Jan. 2013.

[22] C. Damm, M. Schussler, M. Puentes, H. Maune, M. Maasch, and R. Jakoby, "Artificial transmission lines for high sensitive microwave sensors," *IEEE Sensors Conf.*, Christchurch, New Zealand, pp. 755–758, Oct. 2009.

[23] C. Damm, Artificial Transmission Line Structures for Tunable Microwave Components and Microwave Sensors, Shaker Verlag, Aachen, Germany, 2011.

[24] M. Schueler, C. Mandel, M. Puentes, and R. Jakoby, "Metamaterial inspired microwave sensors," *IEEE Microw. Mag.*, vol. 13, no. 2, pp. 57–68, Mar. 2012.

[25] T. Chen, S. Li, and H. Sun, "Metamaterials application in sensing", *Sensors,* vol. 12, pp. 2742-2765, June 2012.

[26] L. Su J. Mata-Contreras, P. Vélez and F. Martín, "Estimation of conductive losses in complementary split ring resonator (CSRR) loading an embedded microstrip line and applications", *IEEE MTT-S Int. Microw. Symp.* (IMS'17), Honolulu, Hawaii, June 2017.

[27] L. Su, J. Naqui, J. Mata-Contreras and F. Martín "Modeling metamaterial transmission lines loaded with pairs of coupled split ring resonators", IEEE Antennas and Wireless Propagation Letters, vol. 14, pp. 68-71, 2015.

[28] J. Naqui, L. Su, J. Mata, F. Martín, "Analysis of transmission lines loaded with pairs of coupled resonant elements and application to sensors", Journal of Magnetism and Magnetic Materials, vol. 385, pp 144-151, June 2015.

[29] L. Su, J. Naqui, J. Mata-Contreras, F. Martín, "Modeling and applications of metamaterial transmission lines loaded with pairs of coupled complementary split ring resonators (CSRRs)", IEEE Antennas and Wireless Propagation Letters, vol. 15, pp. 154-157, 2016.

[30] J. Naqui, C. Damm, A. Wiens, R. Jakoby, L. Su, J. Mata-Contreras, and F. Martín, "Transmission Lines Loaded with Pairs of Stepped Impedance Resonators: Modeling and Application to Differential Permittivity Measurements", IEEE Transactions on Microwave Theory and Techniques, vol. 64, no. 11, pp. 3864-3877, Nov. 2016.

[31] L. Su, J. Mata-Contreras, J. Naqui, and F. Martín, "Splitter/combiner microstrip sections loaded

with pairs of complementary split ring resonators (CSRRs): modeling and optimization for differential sensing applications", IEEE Transactions on Microwave Theory and Techniques, vol. 64(12), pp. 4362-4370, Dec. 2016.

[32] L. Su, J. Mata-Contreras, and F. Martín, "Configurations of Splitter/Combiner Microstrip Sections Loaded with Stepped Impedance Resonators (SIRs) for Sensing Applications", Sensors, vol. 16(12), paper 2195, 2016, doi:10.3390/s16122195.

[33] L. Su, J. Mata-Contreras, P. Vélez and F. Martín, "A Review of Sensing Strategies for Microwave Sensors based on Metamaterial-Inspired Resonators: Dielectric Characterization, Displacement and Angular Velocity Measurements for Health Diagnosis, Telecommunication and Space Applications", International Journal of Antennas and Propagation, vol. 2017, Article ID 5619728, 13 pages, 2017, doi: org/10.1155/2017/5619728.

[34] P. Vélez, L. Su, K. Grenier, J. Mata-Contreras, D. Dubuc, and F. Martín, "Microwave microfluidic sensor based on a microstrip splitter/combiner configuration and split ring resonators (SRR) for dielectric characterization of liquids", IEEE Sensors Journal, published online.

[35] J. Naqui, M. Durán-Sindreu and F. Martín, "Novel sensors based on the symmetry properties of split ring resonators (SRRs)," *Sensors*, vol 11, pp. 7545−7553, 2011.

[36] J. Naqui, *Symmetry properties in transmission lines loaded with electrically small resonators: circuit modeling and applications*, Springer, Heidelberg, Germany, 2016.

[37] J. Naqui, M. Durán-Sindreu, and F. Martín, "Alignment and Position Sensors Based on Split Ring Resonators," *Sensors*, vol. 12, pp. 11790−11797, 2012.

[38] A.K. Horestani, C. Fumeaux, S.F. Al-Sarawi, and D. Abbott, "Displacement sensor based on diamond-shaped tapered split ring resonator," *IEEE Sens. J.*, vol. 13, pp. 1153–1160, 2013.

[39] A.K. Horestani, D. Abbott, and C. Fumeaux, "Rotation sensor based on horn-shaped split ring resonator," *IEEE Sens. J.*, vol. 13, pp. 3014–3015, 2013.

[40] J. Naqui and F. Martín, "Transmission Lines Loaded with Bisymmetric Resonators and Their Application to Angular Displacement and Velocity Sensors," *IEEE Trans. Microw. Theory Techn.*, vol. 61, no. 12, pp. 4700–4713, Dec. 2013.

[41] J. Naqui and F. Martín, "Angular displacement and velocity sensors based on electric-LC (ELC) loaded microstrip lines," *IEEE Sensors J.*, vol. 14, no. 4, pp. 939–940, Apr. 2014.

[42] A.K. Horestani, J. Naqui, D. Abbott, C. Fumeaux, and F. Martín, "Two-dimensional displacement and alignment sensor based on reflection coefficients of open microstrip lines loaded with split ring resonators," *Elec. Lett.*, vol. 50, pp. 620–622, Apr. 2014.

[43] J. Naqui and F. Martín, "Microwave sensors based on symmetry properties of resonator-loaded transmission lines: a review," *Journal of Sensors*, vol. 2015, Article ID 741853, 10 pages, 2015.

[44] J. Naqui, J. Coromina, A. Karami-Horestani, C. Fumeaux, and F. Martín, "Angular displacement and velocity sensors based on coplanar waveguides (CPWs) loaded with S-shaped split ring resonator (S-SRR)," *Sensors*, vol. 15, pp. 9628–9650, 2015.

[45] J. Naqui, F. Martín, "Application of broadside-coupled split ring resonator (BC-SRR) loaded transmission lines to the design of rotary encoders for space applications", *IEEE MTT-S Int. Microw. Symp.* (IMS'16), San Francisco, May 2016.

[46] J. Mata-Contreras, C. Herrojo, and F. Martín, "Application of split ring resonator (SRR) loaded transmission lines to the design of angular displacement and velocity sensors for space applications", *IEEE Trans. Microw. Theory Techn.*, published online.

[47] A. K. Horestani, J. Naqui, Z. Shaterian, D. Abbott, C. Fumeaux, and F. Martín, "Two-Dimensional Alignment and Displacement Sensor based on Movable Broadside-coupled Split Ring Resonators," *Sensors and Actuators A*, vol. 210, pp. 18–24, April 2014.

[48] C. Herrojo, J. Naqui, F. Paredes and F. Martín, "Spectral Signature Barcodes based on S-shaped Split Ring Resonators (S-SRR)", *EPJ Applied Metamaterials*, vol. 3, pp. 1-6, June 2016.

[49] C. Herrojo, J. Naqui, F. Paredes, F. Martín, "Spectral signature barcodes implemented by multi-state multi-resonator circuits for chipless RFID tags", *IEEE MTT-S International Microwave Symposium* (IMS'16), San Francisco, May 2016.

[50] C. Herrojo, F. Paredes, J. Mata-Contreras, S. Zuffanelli and F. Martín, "Multi-state multi-resonator spectral signature barcodes implemented by means of S-shaped Split Ring Resonators (S-SRR)", *IEEE Trans. Microw. Theory Techn.*, vol. 65, no. 7, pp. 2341-2352, July 2017.

[51] C. Herrojo, J. Mata-Contreras, F. Paredes, F. Martín, "Near-Field chipless RFID encoders with sequential bit reading and high data capacity", *IEEE MTT-S Int. Microw. Symp.* (IMS'17), Honolulu, Hawaii, June 2017.

[52] C. Herrojo, J. Mata-Contreras, F. Paredes, F. Martín, "Microwave encoders for chipless RFID and angular velocity sensors based on S-shaped split ring resonators (S-SRRs)", *IEEE Sensors J.*, vol. 17, no. 15, pp. 4805-4813, August 2017.

[53] O. Weiner, "Die theorie des Mischkorpers fur das Feld der statonare Stromung i. die mittelwertsatze fur kraft, polarisation und energie", *Der Abhandlungen der Mathematisch-Physischen Klasse der Konigl. Sachsischen Gesellschaft der Wissenschaften*, vol.32, pp.509-604, 1912.

[54] S. Preradovic and N. C. Karmakar, "Chipless RFID: bar code of the future," *IEEE Microwave Magazine*, vol. 11, pp. 87-97, 2010.

[55] C. S. Hartmann, "A global SAW ID tag with large data capacity," in Proc. of *IEEE Ultrasonics Symposium*, October 2002, vol. 1, pp. 65–69.

[56] A. Chamarti and K. Varahramyan, "Transmission delay line based ID generation circuit for RFID applications," *IEEE Microw. Wireless Compon. Lett.*, vol. 16, pp. 588-590, 2006.

[57] M. Schüßler, C. Damm, and R. Jakoby, "Periodically LC loaded lines for RFID backscatter applications," in Proc. of *Metamaterials 2007*, Rome, Italy, October 2007, pp. 103-106.

[58] N. Saldanha, D.C. Malocha, "Design Parameters for SAW multi-tone frequency coded reflectors" *2007 IEEE Ultrasonics Symp.*, pp. 2087-2090, 2007.

[59] M. Schüßler, C. Damm, M. Maasch, and R. Jakoby, "Performance evaluation of left-handed delay lines for RFID backscatter applications," in Proc. of the *IEEE MTT-S International Microwave Symposium 2008*, pp. 177-180.

[60] S. Harma, V.P. Plessky, C.S. Hartmann, W. Steichen, "Z-path SAW RFID tag" *IEEE Trans. Ultrasonics, Ferroelectric Freq. Control*, vol. 55, pp. 208-213, 2008.

[61] H. Tao, W. Weibiao, W. Haodong, S. Yongan, "Reflection and scattering characteristics of reflectors in SAW tags", *IEEE Trans. Ultrasonics, Ferroelectric Freq. Control*, vol. 55, pp. 1387-1390, 2008.

[62] S. Harma, V.P. Plessky, L. Xianyi, P. Hartogh, "Feasibility of ultra-wideband SAW RFID tags meeting FCC rules" *IEEE Trans. Ultrasonics, Ferroelectric Freq. Control*, vol. 56, pp. 812-820, 2012.

[63] F.J. Herraiz-Martínez, F. Paredes, G. Zamora, F. Martín, and J. Bonache, "Printed magnetoinductive-wave (MIW) delay lines for chipless RFID applications", *IEEE Trans. Ant. Propag.*, vol. 60, pp. 5075-5082, Nov. 2012.

[64] S. Preradovic, I. Balbin, N. C. Karmakar, and G. F. Swiegers, "Multiresonator-based chipless RFID system for low-cost item tracking," *IEEE Trans. Microw. Theory Techn.*, vol. 57, pp. 1411-1419, 2009.

[65] S. Preradovic and N. C. Karmakar, "Design of chipless RFID tag for operation on flexible laminates," *IEEE Anten. Wireless Propag. Lett.*, vol. 9, pp. 207-210, 2010.

[66] O. Rance, R. Siragusa, P. Lemaître-Auger, E. Perret, "Toward RCS magnitude level coding for chipless RFID," *IEEE Trans. Microw. Theory Techn.*, vol. 64, pp. 2315-2325, Jul. 2016.

[67] J. McVay, A. Hoorfar, and N. Engheta, "Space-filling curve RFID tags," in Proc. of *2006 IEEE Radio Wireless Symp.*, pp. 199–202.

[68] I. Jalaly and D. Robertson, "Capacitively-tuned split microstrip resonators for RFID barcodes,"

in Proc. of *European Microwave Conference*, October 2005, vol. 2, pp. 4–7.

[69] H.-S. Jang, W.-G. Lim, K.-S. Oh, S.-M. Moon, and J.-W. Yu, "Design of low-cost chipless system using printable chipless tag with electromagnetic code", *IEEE Microw. Wireless Compon. Lett.*, vol. 20, pp. 640-642, 2010.

[70] A. Vena, E. Perret, and S. Tedjini, "A fully printable chipless RFID tag with detuning correction technique", *IEEE Microw. Wireless Compon. Lett.*, vol. 22(4), pp. 209-211, 2012.

[71] A. Vena, E. Perret, and S. Tedjini, "Design of compact and auto-compensated single-layer chipless RFID tag", *IEEE Trans. Microw. Theory Techn.*, vol. 60(9), pp. 2913-2924, Sept. 2012.

[72] A. Vena, E. Perret, and S. Tedjini, "High-capacity chipless RFID tag insensitive to the polarization", *IEEE Trans. Ant. Propag.*, vol. 60(10), pp. 4509-4515, Oct. 2012.

[73] M. M. Khan, F. A. Tahir, M. F. Farooqui, A. Shamim, H. M. Cheema, "3.56-bits/cm^2 compact inkjet printed and application specific chipless RFID tag," IEEE Ant. Wireless Propag. Lett., vol. 15, pp. 1109-1112, 2016.

[74] M. A. Islam and N. C. Karmakar, "A novel compact printable dual-polarized chipless RFID system," *IEEE Trans. Microw. Theory Techn.*, vol. 60, pp. 2142-2151, Jul. 2012.

[75] R. Rezaiesarlak and M. Manteghi, "Complex-natural-resonance-based design of chipless RFID tag for high-density data," *IEEE Trans. Ant. Propag.*, vol. 62, pp. 898-904, Feb. 2014.

[76] M. Svanda, J. Machac, M. Polivka, J. Havlicek., "A comparison of two ways to reducing the mutual coupling of chipless RFID tag scatterers," in Proc. of *21st International Conference on Microwave, Radar and Wireless Communications (MIKON)*, May 2016, pp. 1-4.

[77] A. Vena, E. Perret, S. Tedjini, "Chipless RFID tag using hybrid coding technique," *IEEE Trans. Microw. Theory Techn.*, vol. 59, pp. 3356-3364, Dec. 2011.

[78] A. Vena, E. Perret, S. Tedjini, "A compact chipless RFID tag using polarization diversity for encoding and sensing", *2012 IEEE Int. Conf. RFID*, pp. 191-197, 2012.

[79] I. Balbin, N.C. Karmakar, "Phase-encoded chipless RFID transponder for large scale low cost applications", *IEEE Microw. Wireless. Comp. Lett.*, vol. 19, pp. 509-511, 2009.

[80] S. Genovesi, F. Costa, A. Monorchio and G. Manara, "Chipless RFID tag exploiting multifrequency delta-phase quantization encoding", *IEEE Ant. Wireless, Propag. Lett.*, vol. 15, pp. 738-741, 2015.

[81] C. Herrojo, J. Mata-Contreras, F. Paredes, Ferran Martín, "Near-field chipless RFID system with high data capacity for security and authentication applications", IEEE Transactions on Microwave Theory and Techniques, submitted.

[82] C. Herrojo, J. Mata-Contreras, F. Paredes, A. Núñez, E. Ramón, F. Martín, "Near-field chipless-RFID tags with sequential bit reading implemented in plastic substrates", Journal of Magnetism and Magnetic Materials, to be published.

Institute of Physics
Dirac House, Temple Back
Bristol BS1 6BE UK

ISSN: 1742-6588
ISBN 978-1-5108-5895-4
9 781510 858954

CURRAN ASSOCIATES INC.
proceedings.com

First Multi-Campus Meeting on Basic Science 2017

Journal of Physics: Conference Series Volume 1002

Medellin, Colombia
29 September 2017

IOP
Institute of Physics